CW01645337

NEBOSH HEALTH AND SAFETY MANAGEMENT FOR CONSTRUCTION (UK)

Unit NC1 Part 1

Contributor

Kevin Coley, MSc, BA, CMIOSH

With thanks to:

Mr Roger Passey, CMIOSH, MIIRSM

Dr T Robson, Bsc (Hons), PhD, CFIOSH, MRSC, CChem

For information on all RRC publications and training courses, visit: www.rrc.co.uk

RRC: Unit NC Part 1

ISBN for this volume: 978-1-912652-39-6
First edition April 2021

ACKNOWLEDGMENTS

RRC International would like to thank the National Examination Board in Occupational Safety and Health (NEBOSH) for their co-operation in allowing us to reproduce extracts from their syllabus guides.

This publication contains public sector information published by the Health and Safety Executive and licensed under the Open Government Licence v.3 (www.nationalarchives.gov.uk/doc/open-government-licence/version/3).

Every effort has been made to trace copyright material and obtain permission to reproduce it. If there are any errors or omissions, RRC would welcome notification so that corrections may be incorporated in future reprints or editions of this material.

Whilst the information in this book is believed to be true and accurate at the date of going to press, neither the author nor the publisher can accept any legal responsibility or liability for any errors or omissions that may be made.

Health and Safety Management for Construction (UK)

Element 1: The Foundations of Construction Health and Safety Management

Element 2: Improving Health and Safety Culture and Assessing Risk

Element 3: Managing Change and Procedures

Element 4: Excavation

Element 5: Demolition

Element 6: Mobile Plant and Vehicles

Element 7: Working at Height

Suggested Answers to Study Questions

Introduction to the Certificate Course

Course Structure

This study text has been designed to provide the reader with the core knowledge needed to successfully complete the NEBOSH Health and Safety Management for Construction, as well as providing a useful overview of health and safety management. It follows the structure and content of the NEBOSH syllabus.

The NEBOSH Health and Safety Management for Construction (UK) consists of 13 elements which are assessed by one exam (Unit NC1). For more detailed information about how the syllabus is structured, visit the NEBOSH website (www.nebosh.org.uk).

NEBOSH Health and Safety Management for Construction (UK)	
Element 1	The Foundations of Construction Health and Safety Management
Element 2	Improving Health and Safety Culture and Assessing Risk
Element 3	Managing Change and Procedures
Element 4	Excavation
Element 5	Demolition
Element 6	Mobile Plant and Vehicles
Element 7	Working at Height
Element 8	Musculoskeletal Health and Load Handling
Element 9	Work Equipment
Element 10	Electricity
Element 11	Fire
Element 12	Chemical and Biological Agents
Element 13	Physical and Psychological Health

The Assessments

Open-Book Exam

The NEBOSH Health and Safety Management for Construction has one assessment - Unit NC1, which is an open-book examination. This new format allows candidates to complete their qualification without needing to attend an exam venue, while maintaining the high standards and robust assessment strategy for which NEBOSH is renowned. This study text contains information and advice on how best to prepare for this exam. NEBOSH has prepared detailed information on the open-book examinations along with a number of useful resources which you can access from the NEBOSH website. You must read and familiarise yourself with the resources NEBOSH has provided; RRC's guidance is complementary, not a replacement for NEBOSH guidance.

MORE...

For the guidance that NEBOSH has prepared see *NEBOSH Open Book Examinations: Learner Guide*, available on the NEBOSH website:

www.nebosh.org.uk

An 'open-book' exam means exactly what the name suggests: you are allowed to access your study text and other materials during the exam. However, it will require more than simply copying text to answer the exam questions; you will need to demonstrate that you understand the subject matter and can apply the topics appropriately - and this needs preparing for. Further guidance can be found in the 'Final Reminders' section later.

The Unit NC1 open-book exam differs from the traditional question-and-answer type format and will instead consist of a realistic workplace scenario which may describe a developing situation such as an incident or safety intervention, and you may be asked to assume a particular role, for example a safety manager. You will then be asked to carry out some tasks and each task will consist of one or more questions, where your answers will mostly be relevant to the scenario.

Another crucial difference between the two exam formats is the duration; since this exam isn't invigilated and requires time to prepare your answers, you have a 48-hour window in which to prepare, complete and submit the exam, therefore allowing for enough time to reflect and detail your answers. It may sound like a lot of time, but there is a lot you will need to do. Don't worry though - we will outline what you need to do in those 48 hours to successfully finish the exam.

NEBOSH specifies 8 hours to complete the exam; the 48 hours allow for you to prepare and choose the best time of day for you to work on the exam - it does not mean you should be working all of that time.

There is a word count currently specified of 4,500 words in total which you would not be expected to exceed. (Note that the word count is subject to change by NEBOSH.) Be aware though that it will be more difficult to pass the exam if your word count is significantly below the specified figure. The word count might provide a guide as to how extensive your answers should be and will help you to focus on what the exam is asking you to do. If we assume a word count of 4,500 words for example, distributed across all the questions, then 10-mark questions require around 450 words while 15-mark questions would require around 500 words. You also have the choice to type your answers or hand-write them.

Following the exam there will be a closing interview - you must take part in this for your mark to be awarded.

In addition to RRC's guidance, you must read the *NEBOSH Open Book Examinations: Learner Guide and the NEBOSH Open Book Examinations: Technical Learner Guide*, which you can download from the NEBOSH website.

MORE...

NEBOSH has prepared detailed information on the open-book examinations along with a number of useful resources which you can access at: www.nebosh.org.uk/open-book-examinations/

HINTS AND TIPS

As you work your way through this book, always remember to relate your own experiences in the workplace to the topics you study. An appreciation of the practical application and significance of health and safety will help you understand the topics.

Learning Outcomes and Assessment Criteria

These learning outcomes and assessment criteria are published by NEBOSH in the syllabus guide for the course. It is very important that you understand all the learning outcomes.

Learning outcome The learner will be able to:	Related content	Assessment criteria	Assessment
Justify health and safety improvements using moral and financial arguments	**1.1**	1.1 Discuss the moral and financial reasons for managing health and safety in the workplace	OBE
Advise on the main roles, competencies and duties under construction legislation and on a range of general construction site issues	**1.2 - 1.7**	1.2 Summarise the main health and safety duties under the Construction (Design and Management) Regulations 2015 and how contractors should be selected, monitored and managed 1.3 Summarise the types of construction work and range of activities 1.4 Identify what to consider during a construction site assessment 1.5 Identify how to keep a site secure and in good order 1.6 Summarise how to manage temporary construction works 1.7 Outline what welfare arrangements should be on site and other particular construction issues	OBE
Positively influence health and safety culture and behaviour to improve performance in their organisation	**2.1 - 2.3**	2.1 Describe the concept of health and safety culture and how it influences performance 2.2 Summarise the human factors which positively or negatively influence behaviour at work in a way that can affect health and safety 2.3 Summarise how health and safety culture at work can be improved	OBE

Learning outcome The learner will be able to:	Related content	Assessment criteria	Assessment
Recognise workplace changes that have significant health and safety impacts and effective ways to minimise those impacts	**3.1**	3.1 Discuss typical workplace changes that have significant health and safety impacts and ways to minimise those impacts	OBE
Develop basic safe systems of work (including taking account of typical emergencies) and knowing when to use permit-to-work systems for special risks	**3.2 - 3.4**	3.2 Describe what to consider when developing and implementing a safe system of work for general activities 3.3 Explain the role, function and operation of a permit-to-work system 3.4 Discuss typical emergency procedures (including training and testing) and how to decide what level of first aid is needed in the workplace	OBE
Take part in incident investigations	**3.5**	3.5 Explain why and how incidents should be investigated, recorded and reported	OBE
Assessing risks, recognising a range of common hazards, evaluating risks (taking account of current controls), recommending further control measures, and planning actions	**2.4** **4 - 13**	2.4 Explain the principles of the risk assessment process 4 - 13 Produce a risk assessment that considers a wide range of identified construction hazards (drawn from elements 4 - 13)	OBE

These learning outcomes and assessment criteria are published by NEBOSH in the syllabus guide for the course.

Keeping Yourself Up to Date

The field of health and safety is constantly evolving and, as such, it will be necessary for you to keep up to date with changing legislation and best practice.

RRC International publishes updates to all its course materials via a quarterly newsletter, which alerts students to key changes in legislation, best practice and other information pertinent to current courses.

Please visit www.rrc.co.uk/news-resources/newsletters.aspx to access these updates.

Other Publications

Study Aids

- NEBOSH National Health and Safety Management for Construction (UK) Revision Guide
- RRC Health and Safety Law Step Notes
- RRC Occupational Health and Hygiene Step Notes
- RRC Quality Management Step Notes
- RRC Environmental Step Notes

Further Your Studies

- NEBOSH National Diploma in Occupational Safety and Health – Unit A: Managing Health and Safety
- NEBOSH National Diploma in Occupational Safety and Health – Unit B: Hazardous Substances/Agents
- NEBOSH National Diploma in Occupational Safety and Health – Unit C: Workplace and Work Equipment Safety

RRC International is continually adding to its range of publications. Visit www.rrc.co.uk/publications.aspx for a full range of current titles.

Element 1

The Foundations of Construction Health and Safety Management

Learning Objectives

Once you've studied this element, you should be able to:

1. Outline the moral and financial consequences of failing to manage health and safety within the construction industry.
2. Outline the scope and application of the **Construction (Design and Management) Regulations 2015**.
3. Understand the type, range and issues relating to construction activities.
4. Describe site assessment and control measures.
5. Outline site order and security, the management of temporary works and other construction issues such as welfare arrangements.

Contents

Morals and Money

IN THIS SECTION...

- There are two main reasons for an organisation to manage health and safety: moral and financial.
- Construction has the largest number of fatal injuries of the main Great Britain (GB) industry groups. Although there is a downward trend in the number of deaths and major injuries, areas of high injury and ill-health incidence remain.

Moral Expectations of Good Standards of Health and Safety

To prevent workplace accidents and illnesses, companies must stop viewing safety simply as complying with the law and start treating it as an ethical or moral issue. Keeping workers safe isn't just about avoiding prosecution; it's about upholding an employer's ethical obligations. Conscientious employers accept that it is an essential part of their corporate social responsibility.

The moral reasons for managing health and safety are based on a person's general duty of care not to cause harm to themselves or others in the workplace. The common law duty of care reflects the moral duty towards the other human who is working for you or with you. Employers and workers must take reasonable care to prevent situations that could cause injury or ill health to others.

The media ensures best practice is known by everyone and establishes a norm that people expect. In recent years, societal attitudes to issues such as drink driving and smoking have prompted changes that have improved standards of health and safety everywhere not just in the workplace. Better staff will only work for employers with better standards.

In simple terms, the moral reason can be summarised as, 'it's the right thing to do'. It is right and proper that workers going to work to earn a living should return home in the same state, not suffering from ill health or serious physical injury. People expect this as a fundamental right. Workers expect it. Society expects it. Over time, this societal expectation has been translated into legal standards. In this way, the moral argument drives legislation.

The Financial Cost of Incidents (Insured and Uninsured Costs)

The financial reasons for health and safety can be difficult to calculate, but incidents cost money to the company (lost production, repair of damage, replacement workers), to the injured person (lost wages, lost opportunities) and to society (emergency services, hospitals).

Companies have gone out of business after major incidents.

Insurance companies often now take considerable interest in health and safety performance and employers who fail to identify hazards and manage risks properly may well find their insurance premiums significantly increased. Following prosecutions, fines imposed by the criminal courts can only be met from the employer's own funds as insurance cover is not possible against criminal penalties.

Manual handling injuries remain a cause for concern for inspectors on construction sites

Although there has been a downward trend in the number of deaths and major injuries in GB's construction industry in the last few years, recent statistics show that:

- 7% of British workers work in construction.
- The fatal injury rate (1.31 per 100,000 workers) is three times the All Industry rate.
- In 2018-19 the number of fatal injuries to workers in construction was 30 (147 across all industries) and 7 members of the public; the lowest number on record for the sector.
- Around 54,000 construction workers in GB sustained an injury at work.
- 37% involved over three days and 28% over seven days absence.

For statistics see: www.hse.gov.uk/statistics/industry/construction.pdf

Good construction health and safety is a large contributor to this fall in the number of injuries, but it must be maintained to ensure the number continues to fall. The hazards and risks of construction activities must be recognised and management systems put in place to eliminate or reduce those risks. Typical areas of high-injury incidence are:

- **Falls from height** - still a prime cause of fatalities and major injuries (around 50% of construction fatalities, 32% of specified injuries and 10% of over-seven-day injuries).
- **Slips, trips and falls occurring on the same level** - still causing a number of over-seven-day injuries. In recent years, they accounted for as much as 29% of specified injuries and 21% of over-seven-day injuries.
- **Being struck by falling/moving objects** - materials and objects dropped from access equipment and buildings - 8% of fatalities and 14% of specified injuries.
- **Manual handling** - lifting and carrying on construction sites - a major cause of lost work days, 7% of specified injuries and 29% of over-seven-day injuries.

Health and Safety Executive (HSE) statistics show that:

- Between 2014 -15 and 2018 -19, each year around 80,000 construction workers in GB suffer from an illness they believe was caused or made worse by their work.
- Around 40% of these cases were new conditions which started during the year, while the remainder were long-standing conditions.
- An estimated 2.3 million working days (full-day equivalent) were lost each year in construction between 2014 -15 and 2018 -19 due to workplace injury (17%) and work-related illness (83%).
- The total cost of workplace injury and new cases of work-related ill health in construction in 2018 -19 is estimated to be £1.2 billion (19% from injury, 81% from illness). This accounts for around 8% of the total cost across all industries, that is, £14.9 billion.
- Assuming a full-time working year equates to 225 working days, this is equivalent to around 10,000 full-time workers being absent from the workforce for the whole year.

TOPIC FOCUS

Despite major advances in health and safety culture and awareness, the construction industry continues to deliver the highest single industry fatality rate in the annual HSE statistics bulletin. Since 1 February 2016, the Sentencing Council has issued guidelines on fines for health and safety offences, including corporate manslaughter, which apply to all sentences and. tThese can be very substantial.

If the court establishes there was a deliberate breach of, or flagrant disregard for, the law, then this is regarded very seriously (high culpability) and large organisations in particular could be presented with an unlimited fine on conviction, or one at least ranging from £500,000 to £10 million.

Even small organisations that are convicted of causing death or injury to their employees by allowing them to working in an unsafe or negligent manner can now be fined up to £1.6 million.

So as you can see, the enforcing agencies and the judiciary are now drawing a clear link between the poor statistics of the construction industry and the punishments awarded. Any organisation, large or small, will, if convicted, now find itself appropriately and proportionately punished.

STUDY QUESTIONS

1. Incidents will have a financial impact that affects the business and can also affect society. Give two examples of costs to:
   (a) The business.
   (b) Society.
2. Identify two risk areas in construction that have a high-injury incidence rate.

(Suggested Answers are at the end.)

The Construction (Design and Management) Regulations 2015

IN THIS SECTION...

- The **CDM Regulations 2015** apply to **all** construction projects in Great Britain.
- Duties are placed upon clients, designers, principal designers, principal contractors and contractors.
- The management of projects can be divided into two distinct stages:
  - Development of pre-construction health and safety information.
  - Development of a construction phase health and safety plan.
- Pre-construction (and other) information must be enough to ensure that all significant risks foreseen in the project are anticipated and planned for.
- The health and safety file becomes the manual for future construction projects and is required for projects involving more than one contractor.
- The extent to which domestic clients must carry out the client duties in **CDM Regulations 2015** is limited and most of the duties are passed to other duty holders.

Roles, Competence and Duties under the CDM Regulations

The **Construction (Design and Management) (CDM) Regulations 2015** provide a detailed framework for managing construction projects. They specifically target planning and design issues, as these areas hold the underlying causes of accidents, as well as the actual construction phase management. The intention is to 'design out' the problems from the start.

CDM Regulations 2015 develop the roles of people involved in construction projects (clients, designers, principal designers, principal contractors and contractors) and places particular duties upon each of them.

In particular:

- The client has a duty to ensure that both the principal designer and principal contractor comply with their duties.
- The principal designer is responsible for managing the pre-construction phase of the project.
- The principal contractor is responsible for managing the construction phase of the project.

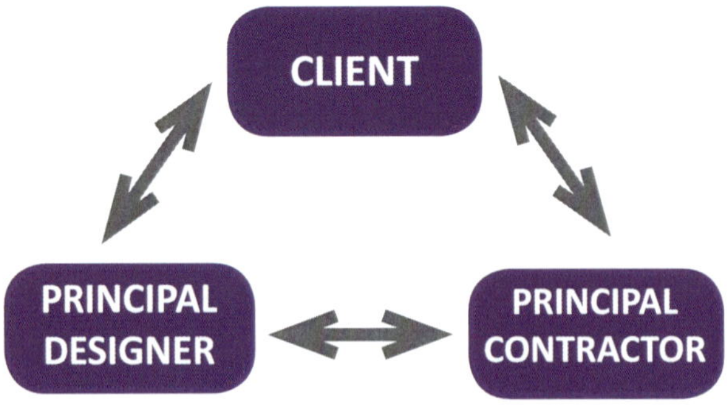

Based on original image from *Industry guidance for principal contractors* (www.citb.co.uk/documents/cdm%20regs/2015/cdm-2015-principal-contractors-printer-friendly.pdf)

They apply to all construction projects, but identify specific types of projects as 'notifiable'.

Clients (Regulations 4, 5 and 6)

Clients must make suitable arrangements for managing a project, including the allocation of sufficient time and other resources, and ensure that these arrangements are maintained and reviewed throughout the project. They must also provide pre-construction information as soon as is practicable to every designer and contractor appointed to the project. In addition, they must ensure that, before the construction phase begins, a construction phase plan is drawn up by the contractor or principal contractor; and that the principal designer prepares a health and safety file for the project.

Designers (Regulation 9)

A designer can be any person who specifies construction work, or specifies materials to be used. Typical examples include architects, surveyors and design and build contractors. The designer must ensure the client is aware of their CDM duties and seek to avoid hazards or minimise risk by effective design. Where risks cannot be avoided, adequate information must be provided with design drawings and specifications.

Principal Designers (Regulation 11)

Principal designers are responsible for planning, managing, monitoring and co-ordinating health and safety in the pre-construction phase of a project. This includes identifying, eliminating or controlling foreseeable risks and ensuring that designers carry out their duties. They are also responsible for preparing and providing relevant information to other duty holders; in particular, principal contractors, to help them plan, manage, monitor and co-ordinate health and safety in the construction phase.

The principal contractor will put the site rules in place

Principal Contractor (Regulation 13 and 14)

The principal contractor is the main contractor involved and is normally a specialist in managing construction projects. Significant to this role is the planning, managing and monitoring of the construction phase, and co-ordinating matters relating to health and safety during the construction phase to ensure that, so far as is reasonably practicable, construction work is carried out without risks to health or safety.

The principal contractor must also consult and engage with the workforce to ensure that measures for their health, safety and welfare are developed, promoted and checked for effectiveness.

TOPIC FOCUS

Principal contractors' duties under the **CDM Regulations 2015** include:

- Preparing the construction phase plan.
- Organising co-operation between contractors (including successive contractors on the same construction site).
- Co-ordinating implementation by the contractors of legal requirements for health and safety.
- Ensuring that employers and self-employed persons apply the general principles of prevention in a consistent manner and follow the construction phase plan.
- Ensuring that a suitable site induction is provided, steps are taken to prevent access by unauthorised persons to the site, and adequate welfare facilities are provided throughout the construction phase.
- Liaising with the client and the principal designer and sharing information relevant to the planning, management and monitoring of the pre-construction phase and the co-ordination of health and safety matters during the pre-construction phase.
- Ensuring workers are being consulted and engaged in securing their health and safety.

Contractors (Regulation 15)

Contractors are those actually carrying out the construction work.

They are responsible for planning, managing and monitoring construction work under their control so that it is carried out without risks to health and safety. For projects involving more than one contractor, they must co-ordinate their activities with others in the project team and, in particular, comply with directions given to them by the principal designer or principal contractor. For single-contractor projects, it is their responsibility to prepare a construction phase plan.

Contractors are also responsible for ensuring that persons they employ to work on a construction site have the necessary skills, knowledge, training and experience to secure their health and safety, and also have appropriate supervision, instructions and information.

TOPIC FOCUS

The competence of contractors is to be assessed by the client. This is usually done using a checklist, covering areas such as:

- Are they experienced in the type of work to be carried out?
- Do they have a suitable health and safety policy in place?
- Do they have suitable arrangements in place?
- What is the quality and extent of their risk assessments?
- What is their recent health and safety performance (number of accidents, etc.)?
- Has any enforcement action been taken against them?
- Do they provide suitable, up-to-date method statements?
- Do they monitor health and safety and carry out site inspections?
- What are the qualifications and skills that they bring to the project?
- Do their employees carry Construction Skills Certification Scheme (CSCS) cards?
- Are they members of a professional body or trade association?
- Do they have employers' liability and public liability insurance?
- Do you have any references about them from previous clients?
- How do they appraise and select subcontractors?
- What are their client liaison arrangements?

The principal contractor may choose not to use a contractor who cannot meet required standards, and exclude them from future lists if their performance is seen to be poor.

Workers

Workers are defined in the **CDM Regulations 2015** as *"...the people who work for or under the control of contractors on a construction site"*.

The duties imposed on workers by the **CDM Regulations 2015** are a repeat of those placed on employees by the **Management of Health and Safety at Work Regulations 1999** to:

- Take care of their own health and safety and others who may be affected by their actions.
- Co-operate with any other person working on, or in relation to, a project at the same or an adjoining construction site.
- Report anything they are aware of in relation to the project which is likely to endanger their own health or safety or that of others.

Domestic Clients (Regulation 7)

Domestic clients are people who have building work carried out in their own home that is not business related. The importance here is the client, not the property. Landlords, charities, housing associations, etc. may own dwellings, but they are NOT domestic clients. Likewise, the owner of a shop being altered that is attached to their home is NOT a domestic client.

The extent to which domestic clients must carry out the client duties in the **CDM Regulations 2015** is limited and most of the duties are passed to other duty holders. A domestic client is not required to carry out the duties placed on commercial clients such as managing projects, notification and the general duties in Regulation 8 of **CDM Regulations 2015**. If the project involves only one contractor, the contractor must carry out the client duties, as well as the duties they already have as contractor, i.e. manage the work to ensure health and safety. If the project involves more than one contractor, the principal contractor must carry out the client duties as well as the duties they already have as principal contractor. If the domestic client has not appointed a principal contractor, the duties of the client must be carried out by the contractor in control of the construction work.

A domestic client has the flexibility of agreeing (in writing) with their designer that the designer co-ordinates and manages the project. Where no such agreement is made, the principal contractor will automatically take over the project management responsibilities.

When the HSE Need to be Notified

The relevant enforcing authority (HSE, Office of Rail Regulation (ORR) or Office for Nuclear Regulation (ONR)), must be informed (using Form F10) if the project is expected to:

- last longer than 30 working days and have more than 20 workers working on the project at any one time;
- or exceed 500 person days.

The client must ensure that an up-to-date copy of the notice submitted is displayed in the construction site office so it is accessible to anyone working on the site, and it must be in a form that can be easily understood. The client can either do this themselves, or ask the principal contractor or contractor to do so.

Form F10

This is the form from the HSE called the 'Notification of a Construction Project' that is completed by the client before a project starts. When completed, it is sent to the HSE, and a copy is displayed on the site. It can be re-sent if further information regarding the project becomes available.

Pre-Selection and Management of Contractors Including Third-Party Auditing Schemes

Before engaging any contractor it is essential to confirm their competence and their procedures that are in place to ensure that the work can be carried out safely. Arrangements will have to be put into place to ensure the effective management of sub-contractors.

Information to prove competence includes:

- health and safety policy;
- suitable and sufficient risk assessments;
- method statements/permit-to-work procedures for dealing with identified hazards;
- suitability of plant and equipment;
- sufficient number of workers allocated to the project;
- past health and safety record;
- checking references with previous clients; and
- membership of trade associations or professional bodies that offer independent assessment of competence such as the Asbestos Removal Contractors Association (ARCA).

A site contact for the contractor should be established before work begins. The site contact will then be responsible for ensuring that:

- regular safety meetings are held;
- health and safety information is communicated;
- the contractor has received and agreed a copy of the site rules;
- agreed working practices, such as permits to work, are being followed;
- regular inspections are completed ;
- any accidents are reported and, if necessary, investigated; and
- a thorough review is conducted on completion of the work.

Effective Planning and Co-ordination of Contracted Work Including Interaction with Existing Staff

The **Management of Health and Safety at Work Regulations 1999** require that, where there is more than one employer working together, that they must co-operate with each other and ensure that procedures are in place to keep other employers and their employees informed about any risks that might arise out of their undertaking.

Co-operation between client, contractor and occupier is very important. The occupier will have a detailed knowledge of any site-specific hazards that may not be obvious to a contractor undertaking construction works and this may impact upon the works. Furthermore, the client or occupier has the authority to place controls and restrictions on the site. Those precautions must take into account those employees who work at the site but are not involved in the construction work. Erecting barriers and warning signs, informing employees about the work or doing the work outside of normal working hours should be part of the planning process.

Co-operation will also be required where site activities need to be controlled or access restricted, or where a shared knowledge is required to undertake a task, for example, decommissioning or removal of machinery.

Health and Safety Planning

There are two distinct stages in the management of a construction project. The first is the development of pre-construction health and safety information which is then developed into a construction phase health and safety plan.

The amount of pre-construction (and other) information that is provided must be enough to ensure that all significant risks foreseen in the project are anticipated and planned for. This information should go beyond what designers and contractors might reasonably be expected to identify, rather than concentrate on the obvious and most likely hazards. As well as those issues mentioned above, it should focus on environmental hazards, restrictions and existing on-site risks, health hazards, design and construction hazards.

The information should be in a convenient form that is clear, concise and easily understood by all parties. There is no reason why existing sources of information (specifications, site plans, etc.) cannot be used, providing they cover the relevant health and safety issues.

The following diagram is taken from Appendix 5 of the guidance on **CDM Regulations 2015** *(Managing health and safety in construction* L153) and summarises how different types of information relate to and influence each other in a construction project.

Start of project

Client to check what documents are already in their possession that will be relevant to the project.

Pre-Construction Information (PCI)

Client and principal designer work together to:

- Assess adequacy of existing information, such as an existing health and safety file.
- Agree arrangements to fill gaps in existing information.
- Provide sufficient information to designers and contractors.

Process of design

Designers must take account of **pre-construction information** to:

- Eliminate, reduce or control foreseeable risks in their designs.
- Provide information to the principal designer about measures taken in designs to reduce or control risks not eliminated.

Principal designer to use this information to:

- Take it into account in the **pre-construction information** and the health and safety file.
- Provide it to the principal contractor towards the construction phase plan.

Construction phase plan

Client to ensure the plan is drawn up before construction phase begins.

Principal contractor to draw up the plan on the basis of:

- **Pre-construction information.**
- Information provided with designs.

Principal designer to help the **principal contractor** prepare the plan.

Principal contractor to:

- Ensure the plan is appropriately reviewed, updated and revised.
- Address any significant changes to risks involved and controls put in place.

Health and safety file

Client to ensure the principal designer prepares the file.

Principal designer to prepare the file in co-operation with the **principal contractor.**

Principal designer to ensure the file is appropriately updated, reviewed and revised.

Principal contractor to provide **principal designer** with relevant information for inclusion in the file.

Principal designer to pass the file to the client at the end of the project.

Principal designer to pass the file to the PC if the **principal designer's** appointment ends before the project finishes.

End of project

Client to:

- Retain the health and safety file.
- Ensure it is available for any subsequent construction work on the building.
- Provide the file to anyone who takes on the client duties (if the client disposes of their interest in the building).

Pre-Construction Information

Pre-construction information provides the health and safety information needed by:

- Designers and contractors to enable them to carry out their duties.
- Principal designers and principal contractors to plan, manage, monitor and co-ordinate the work of the project.

It also provides a basis for the preparation of the construction phase plan and some material may also be relevant to the preparation of the health and safety file. It is information about the project that the client is already aware of or which is reasonably obtainable by the client. However, the information must:

- Be relevant to the particular project.
- Have an appropriate level of detail.
- Be proportionate to the risks involved.

It should be gathered and added to as the design process progresses, and reflect new information about the health and safety risks and how they should be managed. Preliminary information gathered at the start of the project is unlikely to be sufficient.

The information should be presented in a convenient form and be clear, concise and easily understandable to help other duty holders involved in the project to carry out their duties.

TOPIC FOCUS

When pre-construction information is complete, it should include proportionate and relevant information about:

- The project, such as the client brief and key dates of the construction phase.
- The planning and management of the project, such as:
  - Resources and time being allocated to each stage of the project.
  - Arrangements to ensure there is co-operation between duty holders and that the work is co-ordinated.
- The health and safety hazards of the site, including design and construction hazards.
- How health and safety hazards will be addressed.
- Any relevant information in an existing health and safety file.

Construction Phase Plan

The construction phase plan sets out the arrangements for securing health and safety during the period in which construction work is carried out. These arrangements include site rules and any specific measures put in place where work involves particular risks listed in Schedule 3 of **CDM Regulations 2015**.

TOPIC FOCUS

Work involving **particular risks** (Schedule 3 of **CDM Regulations 2015**) includes work:

- Which puts the worker at risk of burial under earth falls or falling from a height.
- Involving dangerous exposure to chemical or biological substances.
- Involving ionising radiation requiring controlled or supervised areas.
- Near high-voltage power lines.
- Which exposes workers to the risk of drowning.
- On wells, underground earthworks and tunnels.
- Involving diving.
- In a compressed-air atmosphere.
- Involving the use of explosives.
- Involving the assembly or dismantling of heavy prefabricated components.

The principal contractor (or contractor in single-contractor projects) must ensure the plan is drawn up during the pre-construction phase and before the construction site is set up. It must take into account the information the principal designer holds, such as the pre-construction information and any information obtained from designers. During the construction phase, the principal contractor/contractor must ensure the plan is appropriately reviewed, updated and revised, so it remains effective.

TOPIC FOCUS

Typically, the construction phase plan will consider:

- A description of the project, such as key dates and details of key members of the project team.
- The management of the work, including:
  - The health and safety aims for the project.
  - The site rules.
  - Arrangements to ensure co-operation between project team members and co-ordination of their work, such as regular site meetings.
  - Arrangements for involving workers.
  - Site induction.
  - Welfare facilities.
  - Fire and emergency procedures.
- The control of any particular risks where they are relevant to the work involved.

Health and Safety File

This document is essentially the manual for future construction projects and is required for projects involving more than one contractor. It contains information about the project to be taken into account when any maintenance or construction work is carried out on the building after the current project has finished. Information included should only be that which is needed to plan and carry out future work safely and without risks to health.

The principal designer must prepare the file, and review, update and revise it as the project progresses. The file must be passed to the client when the project finishes.

TOPIC FOCUS

Typically, the **health and safety file** will contain:

- A brief description of the work carried out.
- Appropriate drawings of structures and layout of services.
- Survey information (e.g. asbestos) and groundwork plans.
- Location of services (water, gas, electricity, etc.).
- Key structural issues and safe loading of floors, roofs, etc.
- Safety information regarding any hazardous substances or materials used.
- Information regarding maintenance issues, e.g. window cleaning.
- Methods of safe access to plant and equipment (e.g. safe access onto roofs).
- Manufacturers' instructions and manuals for new equipment included in the project.

STUDY QUESTIONS

3. Who are the five duty holders under the **CDM Regulations 2015**?
4. When is a project notifiable to the HSE?
5. When do the **CDM Regulations 2015** apply?
6. What documents do the **CDM Regulations 2015** require to be prepared?

(Suggested Answers are at the end.)

Types, Range and Issues Relating to Construction Activities

IN THIS SECTION...

- Construction work involves not just 'building', but also demolition work and building maintenance (occupied and unoccupied).
- Construction work means the carrying out of any building or construction work, including site preparation, erection of new structures, alteration and maintenance of existing structures and other work, including excavation, installation of services, and dismantling or demolition of existing structures.
- These types of work cover a range of activities, including: site clearance; excavation; loading, unloading and storage of materials; movement on site; fabrication; decoration; cleaning; installation, removal and maintenance of service; and landscaping.
- Stability of structure is important to provide adequate health and safety.

Types of Work

DEFINITIONS

STRUCTURE

Any building; timber, masonry, metal or reinforced concrete structure; caisson; mast; or tower.

Structures also include such things as:

- Railway lines and tramways, docks, harbours and canals, airfields, sea defence works, river works, and drainage works including sewers and aqueducts.
- Underground tanks, earth-retaining structures and structures designed to preserve or alter any natural feature, fixed plant and any similar structure.
- Any formwork, falsework, scaffold or other structure designed or used to provide support or means of access during construction work.

EXCAVATION

Work carried out below ground level.

Includes any earthwork, trench work, well, shaft, tunnel or underground working.

A **construction site** is a place where any of the following processes or operations are carried out:

- **Construction**

  This is primarily:

  - The commissioning and erection of new buildings from the foundations and the erection of extensions to buildings.
  - Preparation for intended structures, including site clearance, exploration and investigations.
  - Assembly on site of prefabricated structures, or their disassembly and dismantling.

It includes:

- Excavations and installation of services (e.g. water, sewerage), inland waterways, dams, roads and highways, bridges, tunnels, railways, airports and runways, etc.
- Landscaping of grounds following completion of work or as part of a project.

- **Alteration**

  The modification of the internal and/or external structures and appearance of buildings, including conversion. Also involves 'face-lift' type work, generally improving the appearance of a building without structural works.

- **Maintenance of Premises**

  This is the carrying out of minor repairs, repainting, replacing roof tiles, reglazing windows, etc. in existing occupied and unoccupied premises.

- **Demolition and Dismantling**

  The demolition, dismantling or removal of a structure or part of a structure, or of products and waste is a result of demolition or dismantling of a structure.

- **Site Clearance**

  This involves the removal of hardcore, old building materials and scrap.

- **Excavation**

  Excavation includes earthworks, trenches, wells, shafts, tunnels or underground working.

'Construction work' also includes demolition

- **Structural Work**

  The fabrication, installation or removal of structural parts; decoration, refurbishment or re-decoration after alterations; cleaning of structures using pressure sprays or corrosive substances are all examples of structural work.

- **Site Movements**

  Site movements vary and examples are: loading, unloading, handling and storage of equipment and construction materials; movement of plant, machinery, vehicles and people around the site; use of access equipment, e.g. ladders and stepladders, working platforms, scaffolding, cradles, mobile platforms, crawling boards, gangways, trestles, suspended access equipment, etc.

- **Service Maintenance**

  Installation, removal and maintenance of services include, e.g. electricity, water, gas, compressed air, telecommunications, computer or similar services.

Health and Safety

Regardless of the size of the project, the **construction work** falls into five stages for health and safety purposes:

- **Concept and feasibility** - where the design, layout and construction methods are decided and developed.
- **Design and planning** - shows the specifications required, final drawings and notification to the authorities (where required).
- **Tender/selection** - selection process for the principal contractor (see earlier).
- **Construction phase** - where the construction work is planned, programmed and prepared; then carried out and managed throughout.

- **Commissioning and handover** - all activities involved in bringing the project to an end and the area into operation, e.g. return of services, plant and equipment including the final handover to the client.

DEFINITION

PROJECT

This covers all work on a construction development, including planning, design, management and all construction work until the end of the construction phase.

The Need to Maintain the Stability of Structures

Regulation 19 of **CDM Regulations 2015** deals with the subject stability of structures. Steps must be taken to prevent collapse or instability of structures that may occur during construction work.

Structures must be designed to withstand foreseeable loads and must not be overloaded.

STUDY QUESTIONS

7. What are the five stages of a construction project?
8. What is meant by 'construction work'?
9. What activities take place in construction work?

(Suggested Answers are at the end.)

Site Assessment and Control Measures

IN THIS SECTION...

- Assessment of a site prior to construction is important to identify potential hazards that may be encountered during the work.
- The initial site assessment should look at:
  - The previous and current use of the site and any existing occupied or unoccupied premises.
  - The history of the site, the presence of asbestos or other contaminants, and physical hazards such as underground voids.
  - The area of the site and any restrictions that may apply (special planning requirements or rights of way), the topography and ground conditions.
  - Any non-construction activities that may take place on site.
  - The nature of the surroundings and proximity of features such as roads, footpaths, railways, waterways, etc.
  - The presence of buried or overhead services.
- Construction sites must be well planned to ensure adequate control measures are in place to protect the construction workers, members of the public and vulnerable groups from potential hazards.
- As well as good site planning, consideration should also be given to:
  - Preparation for specialist activities.
  - Site security and protection of the public.
  - Arrangements with clients and occupiers of premises.
  - Arrangements for site inductions.
  - Rules for working in occupied premises.

Initially Assessing the Site

Before starting a construction project, an overall assessment needs to be carried out to identify the significant hazards and physical features which will constrain the site layout, access roads, temporary roads, storage areas, and the siting of office and welfare facilities.

The **CDM Regulations 2015** will apply and the initial site assessment should form the basis for developing the pre-construction information and the construction phase health and safety plan as outlined below.

The initial site assessment should highlight contaminated land issues

The Historical and Current Use of the Land

Other Activities on Site

There may be other activities going on alongside the construction on site, e.g. on shopping streets and in adjacent offices and buildings, that must be taken into consideration. Problems may come from noise, dust, overhead work, vehicle movements, etc.

The Presence of Overhead or Buried Services

DEFINITION

SERVICES

When referring to overhead or buried services, we generally mean: water pipes, gas pipes, oil pipelines, electricity supply cables, sewers and drains, telephone and IT communication cables, etc.

All services within and around the site, both overhead and underground, must be located and identified. Easily missed are services within existing buildings that may remain or be demolished as part of the project.

Likelihood of Asbestos and Contaminants

Asbestos, a naturally occurring fibrous material was used throughout the construction industry for several decades until it was banned in the UK in 1999. This means that many locations that are being developed either through refurbishment of existing premises or demolition and ground clearing activities are likely to encounter asbestos.

Asbestos was usually mixed with other materials, such as cement, it can sometimes be difficult to detect. However, such was the extent of its use that it is reasonable to assume that it is present in buildings that were constructed before 2000. Typical applications of Asbestos Containing Materials (ACMs) are:

- lagging of pipework;
- gaskets;
- water tanks and pipes;
- sprayed coatings; and
- backing panels on gas and electric fires.

Where the land has previously been used before and may still have disused or derelict buildings, there is a significantly increased chance that the land is contaminated. These types of development are often referred to as 'brownfield' sites. Common contaminants are:

- Heavy metals e.g. arsenics, cadmium and lead.
- Oils and tars.
- Chemical substances and preparations, e.g. solvents.
- Gases.
- Asbestos.
- Radioactive substances.

These might be found on brownfield sites which have previously been used as:

- Former factories.
- Mines.
- Steelworks.
- Refineries.
- Landfills.

- Railways.
- Military bases.

Undeveloped land, typically in rural areas where the land may have been used for agriculture or left to evolve naturally, is often referred to as a greenfield site. From a health and safety perspective, it means that there is far less chance of contaminants seeping into the ground and significantly it will mean that there are no buildings to demolish and no industrial roads or debris to remove.

Area of Site

Topography

Topography relates to the study of the surface conditions and shape of the land. A topographical survey of the site will identify all natural and man-made features of the site. This might mean hills, slopes, rivers, streams, etc. together with the height of the land, ground conditions, soil type and other geological features. Each of these features are significant at the initial assessment stage. For example, if the ground is flat and/or has a high-water table, it might be susceptible to flooding; north/east facing slopes will be more susceptible to frost and ice; and, if the rock is granite, there is a possibility that radon gas could be problematic.

When considering construction, it is important to determine whether the area intended as the site has any restrictions placed upon it regarding its development or use. Sites might be within Areas of Outstanding Natural Beauty (AONBs), Sites of Special Scientific Interest (SSSIs), or where preservation orders are in place.

TOPIC FOCUS

Restrictions on land for construction could include:

- National planning authorities placing planning or statutory restrictions.
- Local authorities having regulations, planning restrictions and building by-laws.
- Buildings listed for preservation.
- Government trade and industry regulations requiring permission for industrial development.
- Rights of way.
- Rights of support.
- Rights of light.
- Tunnels, mine workings and abandoned, active or proposed mineral rights.
- Ancient monuments (fortresses, tumuli, etc.) and burial grounds.
- Previous potentially contaminating uses of the site and adjacent areas.

Features of the Surrounding Area

It is important that construction causes no harm to features such as roads, footpaths, railways, waterways, other residential, commercial or industrial properties and schools. Details of the surrounding features before, during and after construction, with drawings and photographs, may be useful.

Site Control Measures

Site Planning

As we have seen, initial site assessment is an important step to ensuring safety during a construction project and providing adequate measures to control people's exposure to the hazards of the site. This not only requires the protection of the construction workers, but also members of the public and vulnerable groups, such as children. Ensuring these controls are adequate relies on good planning and management.

The Need for Safe Entry and Exit

- Access

DEFINITION

ACCESS AND EGRESS

Access is a way in, a way out and a way through a workplace. Egress is a way out. Access and egress can also include ascent and descent.

  - Suitable signs should be placed at site entrances and elsewhere, warning of the site activities.
  - Site traffic will need good visibility entering and leaving the site.
  - Site offices and staff parking should be close to the entrance.
  - Access should be controlled and visitors signed in and out.
  - Arrangements for waste collection and removal should be in place.
  - Access for emergency services should be identified and kept clear.
  - Consideration should be given to any footpaths or public rights of way through the site.
- Roadways
  - All roadways and pedestrian routes should be properly signed.
  - Pedestrians should be segregated from vehicles as much as possible.
  - A one-way system should be put in place, with designated reversing areas if this can't be avoided.
  - Liaison with local authorities may be needed for controls on public roads.
- Storage

  Storage areas should be well signed, secured and protected, and large enough to accommodate what may be left on site. Some materials should be segregated (e.g. gas bottles, chemicals, etc.) and combustibles (timber, etc.) kept away from flammables or heat sources.

- **Loading and Unloading**

DEFINITION

BANKSMAN

A trained person who remains within view of a vehicle driver or crane operator at all times and gives signals to the driver to direct them safely (i.e. hand signals for 'Turn left', 'Slow down', 'Reverse', etc.), at the same time protecting people from the vehicle movement.

  - Safe areas for this will be needed, away from pedestrian activity.
  - A banksman should be used for reversing and unloading operations.
  - Operations may need to be halted while certain crane or unloading activities take place.
- **Offices**
  - Site offices should be as close to the site entrance as possible.
  - They are usually portable buildings and local authorities may need to be notified of their presence.
  - Some buildings may be more than one floor high and adequate fire precautions and escape arrangements must be in place.
  - All temporary services to site offices are to be correctly identified and protected from damage where necessary.
- **Lighting**
  - All site access routes, signs, scaffolding and waste skips should be well lit and free from dark, shadowy areas.
  - There should be enough lighting to allow people to move around the site safely.
  - Some areas, such as pedestrian crossings, may need local lighting.
- **Remediation Works and Provision of Utility Services**

  All areas of site must be safe to work in from the start of the project:
  - Remediation works, such as shoring up unstable structures and covering holes and excavations, may be required. (Note that the term 'remediation' can also refer to works to improve contaminated land.)
  - Existing boundary fences, roadways, etc. may need to be brought up to a safe standard.
  - The site works may need the provision of water, electricity and drainage for construction workers' welfare facilities.

Site Waste Management Plans

Site Waste Management Plans (SWMPs) aim to reduce the amount of waste produced on construction sites and prevent fly-tipping. While there is no longer a legal requirement for construction sites to have SWMPs, it remains good practice to consider the environmental implications of construction work and to responsibly process waste.

Site Preparation for Specialist Activities

Lifting Operations

Using cranes will require good access, firm ground, plenty of room and a safety zone to exclude pedestrians. Overhead safety must also be ensured, to avoid obstructions such as cables.

Piling

This also requires good access and space for the mechanical pile drivers. Noise and dust can be issues here. Safety zones should be applied and all buried services should be identified before any work starts. Cranes may be used together with pile drivers to lift and locate pilings.

Steelwork

Safe areas will be required to bring in and store materials. Cranes may be used to lift the steelwork, so exclusion zones will apply. Safe Systems of Work (SSOWs) for work at height must be in place.

Steel workers require a safe system of work for working at height

Site Security and Client/Occupier Arrangements

Means of Securing Plant and Materials

Construction plant and materials can be both expensive and hazardous, so safe areas are needed to secure such items. This is not only to protect them from damage, but also to protect people on site from the hazards associated with them, particularly children. For example, when chemicals, flammable liquids and gases (such as petrol, diesel and Liquefied Petroleum Gas (LPG)) or hazardous substances (such as pesticides and timber treatment chemicals) are to be stored on site for later use, the principal contractor should ensure a dedicated chemical/hazardous substance store is established (and refuelling area where necessary) - preferably in an area which will not have to be moved for the duration of the project. The area should be protected by heavy-duty barriers, away from all sources of ignition, and from drains and watercourses such as rivers or streams.

All chemicals should be stored in a bund which should be capable of holding 110% of the quantity of the largest container, with all containers clearly labelled with their contents and storage capacity. The chemical store should be locked when not in use to prevent unauthorised access and to reduce the risk of vandalism.

Site Rules

These are usually put in place and enforced by the principal contractor. They are covered in all on-site safety training and displayed at all times in prominent places. Site rules apply to all persons who may come onto the site, not just construction employees, so need to be passed to the occupiers to ensure compliance of their employees and visitors.

TOPIC FOCUS

The principal contractor will put in place rules covering:

- Site access - means of access and authorisations for certain areas.
- Operation of any permit-to-work systems.
- Personal Protective Equipment (PPE) requirements.
- Fire precautions and prevention.
- The nature and accessibility of welfare facilities.
- Site alarms and emergency response.
- First aid and accident reporting.
- Site transport precautions - speed limits, driver rules, parking, reversing, loading and unloading, and deliveries to site.

The type and depth of the rules will depend on the nature and size of the site and the work carried out.

Co-operation and Consultation

Both are vital to ensure that safety covers not only those involved in the construction work, but also persons sharing the site, e.g. occupants of factories, shops, etc. being worked on, or of nearby premises. Separate access and egress for occupants and no-go areas for the construction teams may be necessary.

Shared Facilities

On a dedicated construction site, one set of welfare facilities (toilets, washrooms, rest and eating areas, first aid, etc.) is usually provided and shared by all contractors and participants in the work.

On a site with occupants, there may be facilities already there that can be shared, e.g. toilets, first aid, etc. These arrangements will be agreed and put in place following consultation with the occupiers.

Protection of Other Employees and Visitors

Where occupants of premises being worked on cannot be evacuated for the duration of the work, precautions must be taken to protect them. Methods of separation and segregation may be required. Occupants must be given adequate health and safety information about what is going on around them. On all sites, visitors need particular protection, as they will not be aware of the site layout and the hazards presented by the work being done.

Arrangements for Site Inductions

The principal contractor is responsible for ensuring that site inductions are in place and carried out for all employees and visitors to the site.

All inductions should be recorded and include:

- Details of site arrangements (e.g. signing-in procedures, car parking, use of PPE).
- Conditions of work.

Special hazard signs

- Emergency procedures, including accident reporting, fire precautions, medical welfare where appropriate, locations and purpose of all safety and fire equipment.
- Special hazards and permit-to-work requirements.
- Exclusion and no-go areas.
- Use of mobile telephones, etc.

DEFINITION

TOOLBOX TALKS

These are safety-related briefs, intended to be given to a workforce (by a supervisor) in a short (e.g. ten-minute) talk. They are usually carried out in the immediate workplace. They can cover all topics and records are kept of attendance.

In most cases, site induction will not contain any in-depth health and safety training about specific tasks or hazards, such as work at height or use of plant and equipment. Contractors usually deal with this in a regular series of toolbox talks.

Working in Occupied Premises

Some construction work may take place in occupied premises. This may be in a commercial enterprise, such as an office block or factory, or in a public environment, such as a school, hospital or shopping centre.

We know that safety rules need to be applied to ensure the health and safety of the construction workers, the employees of the client and members of the public. These rules must take into account existing rules that are in place governing the activities of the occupants.

Extra precautions may be required, such as:

- Isolating areas of work from occupants.
- Arranging work out of 'normal' hours (e.g. at night or weekends).
- Appointing wardens to regularly inspect all occupied areas to ensure safety.
- Carrying out more frequent cleaning and waste removal.
- Safety briefs for occupants to make them aware of the hazards and ensure they understand the safety signs, etc.

In all cases, liaison must be ensured and a forum available for questions, comments or complaints to be made by the occupants.

STUDY QUESTIONS

10. What factors should be taken into account in an initial site assessment?
11. List four restrictions that may apply to an area being considered as a construction site.
12. What site security measures may be used on a construction site?
13. What should site rules cover?

(Suggested Answers are at the end.)

Site Order and Security

IN THIS SECTION...

- **CDM Regulations 2015** state that there must be safe access and egress on every construction site.
- Good housekeeping is essential to the safety of a site.
- Secure perimeter fencing:
  - Protects the public.
  - Ensures safe control of children.
  - Should have clear warning signs.
- Arrangements for out-of-hours security such as CCTV and security guards helps to prevent the site from being accessed and damaged.

Site Security

The purpose of site security is to protect both the assets (plant, equipment, chemicals, etc.) on site, and members of the public (especially children) on and off site, who may not be familiar with construction activities.

The **CDM Regulations 2015** address the issues of preventing unauthorised access to the site, good order and site security.

Site order and security consists of:

- Ensuring safe access and egress.
- A programme to ensure good housekeeping.
- Secure perimeter fencing to protect the public and prevent unauthorised access.
- Clear warning signs.
- Out-of-hours security arrangements.

Safe Entry and Exit

Regulation 17 of the **CDM Regulations 2015** states:

> *"There must, so far as is reasonably practicable, be suitable and sufficient safe access to and egress from –*
>
> *(a) every construction site to every other place provided for the use of any person whilst at work; and*
>
> *(b) every place construction work is being carried out to every other place to which workers have access within a construction site."*

Site access and egress must be considered at the design stage. Paragraphs 38 and 39 of the HSE guide *Health and safety in construction* (HSG150) recommend that a traffic management plan should be created which would include how pedestrians can be kept clear during loading and unloading operations, and parking and manoeuvring places where drivers vision might be obstructed. The plan should also address how pedestrians will be segregated from vehicles at the site entrance. This might mean the provision of doors and gates. Gates that open onto traffic routes must have windows.

Access to a construction site should be through a controlled point and requires adequate planning to take into account the surrounding area. Restrictions and hazards relating to safe access to a construction site may include the conditions of the highway from which access is being gained (size, speed, use). The traffic that will operate on the site needs to be considered (size, type, frequency, volume). Traffic routes should be designed so that vehicles do not have to reverse from the site into other traffic.

The site office should be positioned close to the entrance, visitors should be escorted and all deliveries must be signed-in to ensure that only authorised drivers are on site.

Housekeeping Arrangements

Poor housekeeping can lead to incidents such as:

- slips and trips;
- fire;
- biological problems; and
- being struck by falling objects.

Good housekeeping means having in place systems and procedures that will keep the workplace in a clean and orderly condition. In turn, this requires:

- Cleaning up as you go.
- Having a programme of regular waste removal.
- Carrying regular inspections to prevent the build up of waste and debris.
- Having adequate storage facilities by, for example, providing extra bins.

The **CDM Regulations 2015** are very specific. They state that *"no timber or other material with projecting nails (or similar) must be allowed to remain in place if it may be a source of danger"*. The need for good order is further explained in paragraphs 69-73 of HSG150.

Perimeter Fencing

The **CDM Regulations 2015** state that the principal contractor must take the necessary steps to prevent access to the site by unauthorised people. They also state that the perimeter must be clearly identified by signs or fenced off.

All construction sites use security fencing ('Heras' fencing) or hoarding (usually wooden boards) and apply safety signs to warn of the dangers on site.

Fencing should:

- Be at least two metres high and must be well maintained.
- Not be easy to climb.
- Be robust and able to withstand bad weather and extreme environmental conditions. In high winds, hoarding must be able to withstand extreme wind-loading.

If perimeter fencing needs to be removed or altered during the day, it must be replaced by the end of the day and all gates locked securely.

'Heras' site security fencing in easy-to-handle panels
Photo by kind permission of Heras Readyfence

Access should be through gates which can be secured when the site is not in use, and it is good practice to have a security guard at a gate to permit only authorised persons to enter.

Paragraph 40 of HSG150 deals with the subject of site boundaries and paragraph 549 of the same document addresses the subject of protecting members of the public.

Protecting the Public

As well as perimeter fencing options, the following can be used to protect the public:

- Pedestrian tunnels and walkways with lighting.
- False ceilings and crash-decks.
- Sheeting on scaffolding.
- Wrapping protrusions (bolts, couplings, etc.).
- Barriers and signs around excavations.
- Safe viewing points in safe areas and vision panels in solid hoardings.

Precautions for special risk groups, such as disabled persons, should be considered.

Safe Control of Children

Complete perimeter fencing is not always possible, especially on housing sites under development, with coming and going of prospective buyers. This can lead to children trespassing on the site, and site staff need to be aware of this problem, particularly in the evenings, at weekends and during school holidays.

Children must be aware of construction dangers

Visiting local schools, producing pictures and videos, and arranging meetings and talks with parents and their children to make them aware of the hazards and risks on the site can all help to promote site safety awareness and help to reduce accidents with children.

TOPIC FOCUS

In addition to secure fencing of sites, education and publicity, **precautions** that can reduce site risks involving **children** include:

- Removing all ladders from scaffolding, buildings, work platforms, etc. and storing them in a secure area.
- Providing secure and adequate supports around all excavations.
- Immobilising plant, such as dumpers, excavators and other site vehicles.
- Isolating all electrical equipment not required for night safety and security and ensuring the integrity of emergency electricity supplies to the site.
- Reducing the height of all stacked materials.
- Covering any holes that cannot be filled in.
- Blocking off debris chutes by fitting lids or covers to prevent them being used as slides.
- Locking away all chemicals, gas cylinders, tools and equipment in safe and secure storage facilities.
- Providing secure storage tanks for petrol, diesel, oil, etc.

Signs

Signs should be fixed at regular intervals on the perimeter fence to warn of the dangers within the site and instruct people to 'keep out'. When securing signs:

- The health and safety plan will show where on the site layout it is best to place them. Note that they are likely to be moved as work progresses.
- They must be clear and conform to European standards.
- If they are traffic signs, they should be reflective or may require lighting.

Traffic sign indicating lane closures (Source: *Traffic signs manual* (Chapter 8), Department for Transport, 2009 (assets.publishing.service.gov.uk/government/uploads/system/uploads/attachment_data/file/203669/traffic-signs-manual-chapter-08-part-01.pdf))

Out-of-Hours Security Arrangements

When the site is left unattended at weekends, during holidays or after normal work has finished for the day, it is essential that the site is secure. At these times, the main problems that arise are:

- Intruders intending to steal from the site, vandalise equipment or commit arson.
- Trespassers or accidental access to the site.
- Children or young people who see the site as a source of amusement or for criminal activities, e.g. theft or arson.

The following out-of-hours security arrangements may be used:

- **CCTV**

  CCTV may be used to improve site security. To operate within the law, suitable signs and notices must be posted warning of the use of CCTV on the site. When using a CCTV system, signs and notices must be displayed and the requirements of the **Data Protection Act 2018** and the **General Data Protection Regulations** observed.

- **Lighting**

  Suitable site and perimeter lighting can be a useful deterrent.

- **Security Guards**

  It may be necessary to improve the out-of-hours site security by the use of security guards.

- **Storage Compounds**

  Locked and secured compounds for chemicals and tools should be provided.

- **Ladders and Scaffolds**

  Remove ladders from position at the end of the working day or make access to them difficult.

- **Excavations, etc.**

  Barriers and covers should always be used and regularly checked.

- **Falling Objects**

  Steps should be taken to make sure objects cannot fall outside the site boundary by using toe-boards, brick guards and netting on scaffolds.

STUDY QUESTIONS

14. List five ways that the public can be protected from a construction site.
15. What precautions are necessary to prevent accidents to children on a construction site?
16. What six out-of-hours security arrangements might be used on a construction site?

(Suggested Answers are at the end.)

Management of Temporary Works

IN THIS SECTION...

- There are many examples of what temporary works are.
- Temporary works play an important role in every construction or civil engineering project.
- There are reasons that can cause temporary works to fail, including:
  - Overloading.
  - Design fault.
  - Poorly constructed foundation.

Managing Parts of the Works that Allow/Enable Construction of, Protect, Support or Provide Access to, the Permanent Works

Temporary works are any formwork, falsework, scaffold or other structure designed to be used to provide support or means of access during the construction work.

Formwork includes moulds into which concrete can be poured.

Falsework examples include a structure used to support a permanent structure until it has reached a stage where the construction work is sufficiently advanced to support itself. A wooden arch supporting a brick built arch until the cement has set.

Temporary equipment foundations, e.g. the base of a tower crane, are also temporary works.

'Temporary works' are skilfully and deliberately arranged solutions used to:

- Support or protect an existing structure.
- Support or protect permanent works.
- Support an item of plant or equipment.
- Support or protect vertical sides, slopes or stockpiles.
- Provide access.

Examples include:

- Formwork:
  - Moulds into which concrete is poured
- Falsework:
  - A structure used to support a permanent structure until it has reached a stage where the construction work is sufficiently advanced to support itself. A wooden arch supporting a brick-built arch until the cement has set.
- Scaffold.

- Temporary equipment foundations:
  - The base of a tower crane.
  - Anchors and ties for construction hoists.
- Shoring in an excavation.
- Site hoardings and signage.
- Cofferdams and caissons.

When temporary works fail due to overloading, design faults, poorly constructed or inadequate foundation, a structure or building is likely to collapse - often with fatal consequences.

Temporary Works Management

The management of temporary works usually involves temporary works procedures, co-ordination of the activities, appropriate design, through investigation into ground conditions and no unauthorised changes to the design or installation of the temporary works. Managers of temporary works will need to be able to demonstrate that they are competent for the work being undertaken.

Examples of changes that commonly occur, without consulting the temporary works manager include:

- Changing the depth of an excavation.
- Using sheeting on a scaffold which increases the wind loading.
- Advertising material on site hoarding increasing the wind loading.
- Changing the material specified for lower grade materials.

STUDY QUESTION

17. Identify four different types of temporary works.

(Suggested Answer is at the end.)

Other Construction Issues Including Welfare Arrangements

IN THIS SECTION...

- Construction sites must have good standards of welfare facilities, including toilet, washing and rest facilities.
- Temporary Accommodation Units (TAUs) provide accommodation and welfare facilities for site staff and visitors.
- When planning and managing any construction project, issues such as time pressure, inclement weather, use of migrant workers, and literacy and numeracy levels must be taken into account.

Welfare Requirements

MORE...

Further information can be accessed here:

www.hse.gov.uk/pubns/books/l153.htm

Many construction workers are faced with periods away from home, and away from a fixed location with permanent health or welfare facilities. Even so, they should still have the same level of health and welfare provision as at any other place of work.

The specific requirements for health and welfare facilities are found in Schedule 2 of the **CDM Regulations 2015** and HSE publications.

Toilets

Suitable and sufficient sanitary conveniences must be provided or made available. These should be adequately ventilated and lit, and kept clean. Men and women should have separate conveniences, unless they are available in separate lockable rooms which can be used by both.

Washing Facilities

These should consist of washbasins (and showers where health reasons or risks require), and be adjacent to every sanitary convenience and close to every changing room.

Washing facilities must include clean hot (or warm) and cold water, soap and towels (or other means of drying). Washrooms must be suitable, ventilated, adequately lit and kept clean.

Changing Rooms and Lockers

Changing rooms and lockers should be provided or made available where workers need special clothing for their work and cannot reasonably change elsewhere.

Where necessary:

- Men and women are to have separate changing areas.
- Changing rooms are to have seating and clothes-drying facilities.

Welfare facilities include somewhere to change

- Lockable facilities are required for special clothing which is not taken home, clothing not worn at work and personal effects.

Rest and Eating Facilities

Suitable and sufficient rest rooms are required, with separate areas for smokers and non-smokers (e.g. an outdoor smoking shelter). They must have enough tables and seating (with backs) for the numbers likely to be using them at any one time. Separate facilities will be required for new or expectant mothers.

Facilities should:

- Allow meals to be prepared and eaten.
- Provide a means of boiling water.
- Be maintained at an appropriate temperature.
- Be adequately lit and ventilated.

Drinking Water

An adequate supply must be provided or made available at readily accessible and suitable places, conspicuously marked to show it is drinking water, and with cups or other drinking vessels (unless from a drinking fountain).

The Types of Temporary Accommodation Units (TAU) Required for Sites

TAUs, where located in or around a construction or demolition site, provide temporary office accommodation and welfare facilities for site staff. Construction sites will generally require office facilities to provide accommodation for site managers, provide space for meetings and to provide storage for site documentation. Site offices are often described as 'site huts' even if they are large and well fitted out. It is important that site offices are comfortable, attractive and versatile, as well as being suitably robust and secure.

Regulation 17 of the **CDM Regulations 2015** states:

> *"A construction site must, so far as is reasonably practicable, have sufficient working space and be arranged so that it is suitable for any person who is working or who is likely to work there, taking account of any necessary work equipment likely to be used there."*

Requirements of Location for TAU

They are often located along a safe access route from outside of the site enclosure in order to facilitate safe visitor access. However, they can be located within existing buildings, or use part of an existing building as the accommodation itself.

The HSE guide *Fire safety in construction* (HSG168) provides guidance *"for clients, designers and those managing and carrying out construction work involving significant fire risks."* It includes guidance on the use of temporary site accommodation units (such as offices, canteens and welfare facilities.)

The HSE guidance states:

- Temporary site accommodation should be located no less than 6m away from the building work in the open air. If units have to be located closer, the unit or the part of the building adjacent to it, should be fire-resisting.
- On timber-frame and high-risk sites (an initial assessment should establish whether the site is 'high risk'), temporary site accommodation should be separated from the building under construction by 20m to establish a fire break. If this is not possible, other measures must be put in place to make sure the risk is controlled.

Particular Construction Issues

There are a range of considerations which apply to construction sites and activities. As such, the following must be taken into account when planning and managing any construction project.

Use of Migrant Workers

The construction industry relies heavily on migrant workers to ensure that there is always a plentiful supply of skilled workers. However, there are certain issues that employers should consider when taking on workers from abroad:

- Training programmes may need to be adapted.
- They may be motivated to earn as much money as possible and consequently be prepared to take more risks.
- There may be a lack of knowledge regarding health and safety rights and how to raise concerns.
- They may have a limited knowledge of UK health and safety systems.
- They may come from a background which does not share the same health and safety culture as the UK.

Temporary Nature of Construction Activities and the Constantly Changing Workplace

The changing range of activities; the number of different tasks taking place at the same time; a changing workforce with different skills, knowledge, experience, competencies and training - all contribute to an ever-changing situation during construction operations.

Time Pressures from Clients and Principal Contractors

Clients and principal contractors (see earlier) may often impose time pressures on the construction phase of the work. This results in pressure on the various levels of contractors and subcontractors involved in the project, who may be tempted to cut corners rather than complete the work safely and satisfactorily.

Weather Conditions

External work is often carried out in poor weather conditions and may result in diminished safety for the workers.

Levels of Numeracy and Literacy of Workers

Poor literacy and numeracy may be an issue as:

- Workers are likely to be from a wide range of socioeconomic backgrounds with varying levels of education.
- Workers with learning difficulties are employed in the construction industry.
- Written instructions will be a problem to workers suffering from dyslexia, who may nevertheless be adequately skilled in the particular type of work.

Non-English Speaking Workers

Construction workers may be drawn from a wide variety of British nationals, workers from the EU and other overseas countries. This can cause communication difficulties due to English not being the worker's first language. Managers need to be aware of this problem and consider it when assessing risk.

Managers will need to keep all of these factors in mind when making any decisions about the work, particularly where health and safety will be impacted.

STUDY QUESTIONS

18. What welfare facilities does an employer have to provide on a construction site?
19. Identify the issues that need to be considered when planning and managing a construction project.

(Suggested Answers are at the end.)

Summary

This element has introduced you to the general principles of construction law and management which relate to the construction industry.

In particular, this element has:

- Outlined the moral and financial consequences of failing to manage health and safety properly within the construction industry.
- Given statistics showing how large the problem is.
- Explained that good construction health and safety relies on good management.
- Identified the scope and application of the main legal instrument in the construction industry, the **Construction (Design and Management) Regulations 2015**.
- Explained how duties are placed upon persons in control of construction projects to ensure competence, adequate resources and information, and good management from the inception of the project to completion and handover.
- Outlined the type, range and issues relating to construction activities.
- Outlined the main factors to be considered during the initial site assessment including:
  - the historical use of the site (issues such as asbestos, contaminants and greenfield and brownfield sites).
  - Restrictions on carrying out construction.
  - The importance of topography.
- Identified site-appropriate control measures, including site planning, waste management, and preparation for specialist activities (lifting, piling, steelworks, etc.); site security and arrangements with clients and occupiers of the premises.
- Outlined the purpose of site order and security including the need for safe access and egress; housekeeping arrangements; perimeter fencing and out-of-hours security.
- Outlined the purpose, types and management of temporary works.
- Identified health and welfare requirements, including:
  - Toilets and washing facilities.
  - Changing rooms lockers.
  - Rest and eating facilities, etc.
  - Drinking water.
  - The requirements for the provision of TAUs.
- Considered those issues that are particular to the construction industry, including:
  - Migrant workers.
  - Temporary nature of construction activities.
  - Time pressures.
  - Weather conditions.
  - Levels of numeracy and literacy of workers.
  - Non-English-speaking workers.

Exam Skills

Introduction

To pass the NEBOSH Construction Certificate, you need to perform well during your 'open book' exam. An 'open book' exam means exactly what the name suggests: you can access your study text and other materials during the exam. However, it will require more than simply copying text to answer the exam questions; you will need to demonstrate that you understand the subject matter and can apply the topics appropriately – these needs preparing for.

Here we will consider some practical guidelines that can be used to increase success in your 'open book' exam. The NEBOSH Construction Certificate 'open book' exam differs from the traditional question-and-answer-type format and will instead consist of a realistic workplace scenario which may describe a developing situation such as an incident or safety intervention, and you may be asked to assume a particular role, for example a safety manager. You will then be asked to carry out some tasks. Each task will consist of one or more questions, where your answers will be relevant to the scenario.

Wider Reading, Research and References

During the 'open-book' exam you are allowed to access many different types of resources such as RRC study text and revision guides, professional journals, digital resources on the Internet such as HSE guidance, bilingual dictionaries, etc. Remember though that this is not a substitute for thorough study of your course materials and specific preparation in the form of revision to ensure that you are familiar with all the topic areas included in the syllabus. NEBOSH "will not be looking for anything from your answers that has not been covered in the Unit syllabus".

NEBOSH has designed the 'open-book' exam "to assess the same learning outcomes to the same level" as the previous method of examination. It will "measure depth of understanding" and require you to apply your knowledge to the scenario given rather than simply recalling memorised information. It is therefore "very important that you prepare and revise" for the assessment.

Wider Reading and Research

As part of your preparation, NEBOSH recommends that you should:

- "Complete wider reading so you don't have to do this during the examination."
- Conduct "wider research to understand how your studies relate to the real world".

To help with this you should make use of the **MORE...** boxes included in your course material. These highlight additional resources related to the topics you've been studying. Reading them will assist you in increasing the depth of your understanding and applying it to real life situations.

The links in **MORE...** boxes are current at the time of publication of your materials but may have changed since, e.g. due to an item being moved elsewhere, being replaced or perhaps withdrawn from publication altogether.

HINTS AND TIPS

If the link in a **MORE...** box no longer works, you can search for the item in question by entering its title in a search engine such as Google or Bing. This should lead you to its current location if it's still available on the Web. In the event that it's no longer available, your tutor may be able to suggest a replacement.

If a link to a section of the HSE website no longer works, you may be able to find a current link by searching the HSE website. (The A-Z topic list at the foot of the screen is a useful starting point.)

Your wider reading may also include research using other resources mentioned in your course materials or resources that you've found yourself, perhaps by entering a topic name in a search engine. Always be careful though to use only authoritative sources for your research, such as Government or official websites, and to safeguard the security of your personal information.

Making Use of Your Resources

NEBOSH states that "You are expected to offer your own analysis and presentation of information gained from your research". So you need to:

- Organise your notes in advance so that you can find things quickly.
- Use your own words to express what you've learned and apply the knowledge you've gained.

You are permitted though to make reference to sources, ideas or work of other people or organisations. Where you do so, it's important to acknowledge where that information comes from because your open-book exam response must be your own work and you don't want to be guilty of plagiarism and subject to investigation by NEBOSH.

DEFINITION

PLAGIARISE

"To steal ideas or writings from another person and present them as one's own".

(Source: The Chambers Dictionary)

References

Direct Citations

You may want to quote specifically from a publication you've accessed. Here you should put the quote in inverted commas and make it clear where it's from. (See also **Reference List** below.)

Example:

HSE guidance INDG163 states that "A risk assessment is not about creating huge amounts of paperwork, but rather about identifying sensible measures to control the risks in your workplace".

Indirect Citations

You may just want to refer to information that you've read without quoting from it directly. In this case, express it in your own words but make sure that you identify the source. (See also **Reference List** below.)

Example:

HSE guidance (INDG163) identifies five steps to risk assessment.

Reference List

Your exam answers should include a list of the materials you've referred to – those you've quoted from, those you've mentioned and any other materials that you looked at during the exam. This doesn't count towards the word limit for your answers and it would be best to make the list as you write your answers to ensure that the list is complete.

We don't recommend using the Harvard (or any other) referencing system for this purpose because NEBOSH has made it clear that a simple list is all that is required.

Your list should be at the end of your answer document and should include:

- the title of the publication/document,
- the author,
- the year of issue, and
- a Web address (for Internet publications).

Example:

HSG65 *Managing for health and safety*, HSE, 2013

(https://www.hse.gov.uk/pubns/priced/hsg65.pdf)

INDG163 *Risk assessment – A brief guide to controlling risks in the workplace*, HSE, 2014

(https://www.hse.gov.uk/pubns/indg163.pdf)

(In the above Example:

- INDG163 is included because quoted from and referred to in the text.
- HSG65 is included because consulted during the exam.)

Note: If you use the NEBOSH answer sheet, there is a place there for your references.

Exam Technique

The 'open-book' exam will test you on your ability to "demonstrate analytical, evaluation and creative skills as well as critical thinking" and how you apply your learning to your answers. In other words, you will need to show what you can do with your knowledge to solve the problems presented to you – and this may take practice. To assist you in showing your knowledge, let us look at a step-by-step approach that you can adopt when answering your exam questions.

Step 1 **Read the scenario** - the first step is to read the scenario carefully. Take care with this as it is very easy to misread words in the rush to get writing. It's likely you will read the scenario more than once.

Step 2 **Look at the first task** - a task is an activity or piece of work that will be part of a larger project. The task may be split up into several sub-tasks. The task is used to indicate the questions. Read each question carefully. This is what your knowledge is being tested on - so your answer must address the question. Demonstrating knowledge alone will not gain marks - your knowledge has to be applied to the task - so you need to clearly understand the question being asked.

Step 3 **Look at the marks** - each task or sub-task, question or part of a question will have the maximum number of marks indicated in brackets. For each mark to be awarded, the examiner will expect a piece of information that demonstrates an analytical evaluation of the task that has been set. The marks available give an indication of how much you will need to write and, to a lesser extent, how long you should spend on this part of each question.

Step 4 **Re-read** the scenario and task question - to check that you have properly interpreted them and understood them. There are no marks available for answering the task question that you **think** you see rather than the one that the examiner asked you.

Step 5 **Draw up a plan** - this can take the form of a list or a mind map that helps you unload information quickly and make sure you have enough factors (or things) in your answer to gain the available marks. Jotting down a plan can help you remember key points. The plan is also your aide-mémoire to keep you on track as you start to write your full answer. Your plan can use information from the scenario to support your answer.

When it comes to the exam, there is no need to submit your plan.

NEBOSH has said that "You can expect to see questions that ask What? Could? How? Why? and Where? " but "Whatever format the question, it will relate to the syllabus and learning outcomes of your qualification".

When writing your answer, you must ensure that the structure of the task appears in the structure of your answer. So, for example, if the task question has a part (a) and a part (b), your answer must follow the same structure. Answer part (a) and label it clearly for the examiner as the answer to part (a). Then leave a gap (one line will do) and answer part (b) and label it clearly. The examiner must be able to see the two separate parts of your answer and it must be clear to them which parts are the answer to which questions. One long paragraph of text that contains all parts of the answer jumbled up together cannot gain full marks, even if all of the relevant information is there.

The above exam technique is tried and tested and is the best way to approach each exam task. 'Open-book' exams are a new approach for NEBOSH but the education sector has used 'open-book' exams for a long time.

NEBOSH gives a 48-hour window of time for the exam to reflect different time zones learners work and live with. You are not expected to spend 48 hours completing the exam. You will, however, need to monitor your progress on completing all the tasks and you should plan for drafting answers, reflection and amending answers, building in time for breaks.

Remember, too, that there will be an overall word count for the exam. It's important not to exceed the recommended total by more than 10% so do allow time to check this.

Practice Exam Questions

At the end of Elements 1 to 13 there is a practice exam question (or two) for you to attempt, with guidance on how to answer in addition to a suggested answer. This includes an Answer Outline – all of the points listed in this would attract marks and you will see most of them developed in the suggested answer itself.

Remember that when answering exam questions, information from additional reading and personal experience can be included. Examining bodies encourage this and it will enhance your answers.

The study text provides some useful links to external sources - look out for the '**More...**' boxes within the materials - these contain useful links to relevant topics.

Please feel free to contact your tutor if you have any queries or need any additional guidance.

Question

Scenario

You have been appointed as the competent person for health and safety to assist the site manager who is classed as the 'principal contractor' on a new development involving the construction of a multi-story building. The site manager is currently reviewing the work that will be required to be conducted during the construction phase of the project and is drawing up a list of activities that will be subcontracted out. He is aware, as the 'principal contractor' on this project, he is responsible for all activities that go on. He wants to ensure all the work is done safely by all those involved. The 'client' has approached him about the selection of contractors and is unsure what could be used as a measure to ensure the suitability and health and safety competence of potential contractors. A meeting has been scheduled for one hour with yourself, the manager and the client for you to discuss the client's concerns.

Task: CDM Requirements

Prepare some notes for discussion at the meeting. You think 5 minutes should be sufficient for reminding the site manager and client of their legal responsibilities and 10 minutes to cover the contents of a checklist that could be used to assess the health and safety competence of a contractor:

(a) Legal responsibilities. **(2 marks)**

(b) Contents of a checklist. **(8 marks)**

(Total: 10 marks)

Approaching the Question

Now think about the steps you would take to answer this question:

Step 1 The first step is to **read the scenario** carefully. Note that this is a new manager who has become confused with research undertaken. This means information should be easy to understand and examples can be used to support that understanding. We do not want the manager even more confused.

You are asked to provide "some notes" so will need to structure your approach – a bullet-point list would only add to the manager's confusion.

Step 2 Now look at the **task** – prepare some notes using three headings:

(a) Legal responsibilities.

(b) Key points to establish competence from subcontractors.

Step 3 Next, consider the **marks** available. In this task, there are 10 marks distributed between three parts. Tasks that are multi-part are often easier to answer because there are additional signposts in the question to keep you on track. In this task, you have to create some notes that are easy to understand, giving examples for each part that can aid understanding. You will need to provide around 10 or more different pieces of information (including examples) in total for this task. The headings will allow you to keep your responses separate – this will also help the examiner when marking!

Step 4 **Read the scenario and task again** to make sure you understand them and have a clear understanding of 'so far as is reasonably practicable', health and safety regulations and Approved Codes of Practice. (Re-read your study text if you need to.)

Step 5 The next stage is to **develop a plan** – there are various ways to do this. Remind yourself, first of all, that you need to be thinking about 'so far as is reasonably practicable', the purpose and legal status of legislation and an ACoP. Each part requires examples.

Suggested Answer Outline

(a) Legal requirements:

- Client's duties under CDM (Regulation 4, 5 and 6).
- The client is responsible for assessing the competency of contractors.
- This is done using a checklist.
- Principal contractor's duties under CDM (Regulations 13 and 14).
- Responsible for co-ordinating matters relating to health and safety during the construction phase.

(b) Competence of contractors criteria:

- The experience they have had in this type of work.
- Their health and safety policies and practices.
- Their monitoring arrangements.
- The quality of their risk assessments.
- Their accident records.
- Any recent claims or prosecutions.
- The suitability of their method statements.
- What qualifications and skills they have and what qualifications their employees have.

Now have a go at the question yourself.

Example of How the Question Could be Answered

(a) *The dutyholders (the client and the principal contractor) both have responsibility for the management of contractors on site. First, the client has a responsibility to ensure the competency of contractors before they are appointed. This could be carried out by pre-selection screening.*

The principal contractor has a responsibility for co-ordinating matters relating to health and safety during the construction phase.

The duties are mentioned under the Construction Design and Management Regulations 2015 (CDM) Regs.

(b) *To ensure the competency of potential contractors, the client can check relevant documentation such as past experience they have had in completing the type of work required with customer testimonials, in addition to looking at the contractor's health and safety policy along with the quality of risk assessments and method statements. The same would apply to scrutiny of proposed monitoring arrangements, such as site inspections and their frequency, along with the reporting of incidents and accidents, and RIDDOR reporting should this be required. In addition, checking of the contractor's employee's competency in relation to the skills they hold. Proof of relevant training and qualifications would be required along with any refresher training in order to ensure they remain current with their respective disciplines.*

Finally, evidence of past accident history and any criminal and civil litigation they may have incurred. Criminal litigation in the form of enforcement notices and prosecutions can be cross-referenced on the HSE website to ensure legitimacy.

Reasons for Poor Marks Achieved by Exam Candidates

- In part (a) going into all the duties of the client and the principal contractor opposed to narrowing it down to the specific duties regarding contractor selection and management.
- Citing incorrect examples in part b) of the question and providing sufficient detail to each example.
- Exceeding the overall word count for the whole exam by more than 10%. The examiner may not read excessively long explanations.
- Not providing an answer to the task question at all. If you do not attempt all tasks, you cannot get any marks for the task questions you omit.
- Not considering the scenario – in this case a site manager and the client – so overly complex explanations or highly detailed examples may not attract marks.

References

NEBOSH Open Book Examinations: Learner Guide - Guidance document for preparing for the open book examination for NG1, IG1, NGC1 and IGC1 units, NEBOSH, 2020

https://www.nebosh.org.uk/documents/open-book-examination-learner-guide

The Chambers Dictionary, Chambers, 2005

INDG163 *Risk assessment – A brief guide to controlling risks in the workplace*, HSE, 2014

(https://www.hse.gov.uk/pubns/indg163.pdf)

HSG65 *Managing for health and safety*, HSE, 2013

(https://www.hse.gov.uk/pubns/priced/hsg65.pdf)

Element 2

Improving Health and Safety Culture and Assessing Risk

Learning Objectives

Once you've studied this element, you should be able to:

1. Explain health and safety culture and how it affects performance.
2. Explain how human factors influence behaviour.
3. Identify how to improve health and safety culture in organisations.
4. Understand risk assessment.

Contents

Health and Safety Culture

IN THIS SECTION...

- The health and safety culture of an organisation is the way that all the people within the organisation think and feel about health and safety and how this translates into behaviour. It can be defined as the shared attitudes, values, beliefs and behaviours relating to health and safety.
- There is a strong link between health and safety culture and health and safety performance. Organisations with a strong, positive culture tend to have good performance, whereas those with a weak, negative culture perform poorly.
- Workers are often influenced by their peers - the people around them at work who do not have any direct authority over them. This 'peer group pressure' occurs indirectly by social interaction and can have a significant effect on behaviour. The influence of peer group pressure is a good indicator of health and safety culture.

Meaning of the Term 'Health and Safety Culture'

All organisations have a 'culture'. It is not written down, or even easily stated. It is a subtle mix of formal and informal rules, relationships, values, customs, etc., which, taken together, describe the distinctive 'feel' of the organisation. On one level, this is to do with how the organisation gets things done - its particular way of working. On another level, it is to do with how people perceive the organisation, e.g. how friendly it is.

All organisations have a 'culture'

Organisational culture is a characteristic of the organisation that exists at every level, from senior management to production workers. No one person determines the culture of the organisation; all staff working for the organisation determine it collectively.

Organisations can be described as having a 'health and safety culture' (or safety culture) in much the same way.

DEFINITION

HEALTH AND SAFETY CULTURE

The shared attitudes, values, beliefs and behaviours relating to health and safety.

The safety culture of an organisation is the way that everyone within the organisation thinks and feels about health and safety and how this translates into their behaviour.

Relationship Between Culture and Performance

An organisation will have either a positive safety culture or a negative one.

Positive Culture

In an organisation with a **positive** safety culture, the majority of the workers think and feel that health and safety is important. There is a strong policy and clear leadership from the top because senior management has this attitude, which runs through the whole organisation, from top to bottom. Managers think about the health and safety implications of their decisions and workers share the same view and work safely.

Everyone works safely because they want to. That is the way that things are done in their organisation and that is how everybody else is behaving, too.

In People in the organisation who do not share this view are in the minority and are likely to come round to the group way of thinking and acting. This is because the culture of an organisation tends to be absorbed by its workers over time. Workers who do not adjust to the group way of thinking may either leave, because they don't feel that they fit in, or possibly be dismissed for working unsafely.

In a company with a positive safety culture - all members of staff appreciate the importance of safety

In an organisation like this, you can see that there is a clear link between safety culture and health and safety performance. People work safely, so there will be fewer accidents and less ill health. It is also easy to see why organisations strive to create a strong, positive safety culture because when there is one, it has a direct influence on worker behaviour.

Negative Culture

In an organisation with a weak, **negative** safety culture, the majority of workers think and feel that health and safety is not important; they are poorly educated in health and safety and see it as unnecessary or not important. There is a lack of clear direction and leadership from senior management. Managers do not think about health and safety in their decision-making and so let other priorities, such as productivity, dictate their actions. Workers behave unsafely, often because they do not know any better.

Safety-conscious workers are in the minority and are likely to come round to the group way of thinking and acting over time; if not, they may well leave because they do not like the organisational culture and feel unsafe in the work situation.

A workplace with a negative safety culture results in unsafe behaviour

You can see that in an organisation like this there will be a lack of attention to health and safety, standards will be low, behaviour will be poor and accidents may occur as a result.

MORE...

Further information about safety culture can be found on the HSE website at:

www.hse.gov.uk/humanfactors/topics/culture.htm

Influence of Peers

Peer group pressure can influence the thinking about safety-related behaviour

When people are put together into groups, they interact. Some individuals will come to have a lot of influence over the group; others will have little influence. In this way, a 'hierarchy' (often known as a 'pecking order') is established within the group. Certain ways of behaving will become the 'norm', which will often be established by the more influential members of the group. A person wishing to become a member of the group will have to comply with the group norms. This pressure to comply with group norms is 'peer group pressure'.

Peer group pressure is an important factor to take into account when thinking about safety-related behaviour. If a group is already working safely, then peer group pressure will keep most people in that group in line. But if the group is working unsafely, then peer group pressure will tend to force more and more workers to behave unsafely in an attempt to fit in. Even though workers may know that what they are doing is wrong and may want to do it the right way, the pressure to comply with the group overcomes their personal apprehensions.

The way to deal with the negative impact of peer group pressure is usually to tackle the influential people within the group, who are the ones responsible for establishing group behaviour. If their behaviour can be changed, then everyone else's will change as well. This might be done by training, education, involvement in safety-related projects, etc. One very successful tactic is to give them increased responsibility. Ultimately, if the influential members will not change their behaviour then they may have to be moved into other work groups where they may have less influence, or they may have to be disciplined using the normal disciplinary process.

The behaviour of peer groups and the influence that peer group pressure has been allowed to exert on worker behaviour is often a good indicator of safety culture. In an organisation with a positive culture, peer group pressure is in line with safe behaviour. In an organisation with a negative culture, peer group pressure is driving unsafe behaviours and this has not been challenged by management.

STUDY QUESTIONS

1. Define 'health and safety culture'.
2. How do an individual's peers exert influence over their behaviour?

(Suggested Answers are at the end.)

How Human Factors Influence Behaviour Positively or Negatively

IN THIS SECTION...

- Of critical importance to health and safety management is individual worker behaviour. One worker may behave in an ideal manner, but another may not, and this unsafe behaviour may endanger themselves and others.
- Three significant factors influence a worker's behaviour:
  - The organisation - characteristics of the organisation that they are working for.
  - The job - the task that they are carrying out.
  - The individual - their personal characteristics.
- Key characteristics of an individual worker that influence their safety-related behaviour include their attitude (how they think about a particular safety issue) and their ability to perceive the hazards and risks present in their workplace.

Introduction

One issue of critical importance to health and safety management is the way that individual workers behave. It is estimated that well over half of all workplace accidents are caused by unsafe acts: the poor safety-related behaviour of a worker. It is not enough to dismiss this as being due to carelessness; this simply blames the worker and is ineffective at identifying underlying causes or corrective actions. Instead, we must look at how 'human factors' influence working practices; we have to understand why people behave the way they do at work.

If we can understand that, then it may be possible to:

- Correct poor behaviour when it is identified by removing the cause of that behaviour.
- Anticipate poor behaviour before it occurs and introduce changes to reduce the likelihood of it occurring.

Why is it that a worker behaves poorly when working for one organisation, but then leaves and starts to work for another company and behaves in an entirely different manner?

Why is it that a worker may behave safely doing one job, but then unsafe practices start to creep into their behaviour when they are switched to another job?

Why is it that one worker behaves safely at work, but another does not, even though working conditions for both workers are the same?

The answer to these three questions (and the link between them) is 'human factors'. This phrase refers to a range of issues that influence a person's safety-related behaviour when they are at work.

These issues can be grouped under three main headings:

- **Organisational factors** - characteristics of the organisation that they are working for.
- **Job factors** - characteristics of the job or task that they are performing.
- **Individual factors** - characteristics of the individual.

Factors which influence behaviour

Collectively, these are known as the three **human factors** in health and safety - they are the three factors that influence a person's behaviour at work. No one single factor dictates behaviour; they subtly nudge behaviour in one direction or another.

Organisational Factors

These are the characteristics of the organisation that influence workers' behaviour.

Organisational factors include the following features.

Safety Culture of the Organisation

The attitudes, values and beliefs of workers is shown by their inclination to comply with rules or act safely or unsafely. For example, if a company is unsuccessfully relying on procedural controls to avoid major accidents, additional alternative safeguards through the hierarchy of control may need to be applied to ensure safety.

Leadership from Management

This is visibly demonstrated outside of the boardroom (since behaviour in the boardroom is not witnessed by most of the workers in an organisation). Leadership from management can be demonstrated in a number of ways:

- leading by example;
- appointing competent persons;
- appointing a senior manager with health and safety responsibilities;
- being involved in safety committees;
- involvement in active monitoring such as safety tours;
- responding positively to audit/inspection reports;
- ensuring a board-level review of performance;
- consulting with workers;
- establishing a robust health and safety policy;
- dealing promptly and effectively with safety-related issues.

Organisational factors include the allocation of resources - whether financial or provision of adequate time and personnel to carry out tasks safely.

Work Patterns

Patterns of work such as shift systems, work at night or extended hours that can lead to adverse effects on health and fatigue, that in turn can lead to poor performance on tasks that require attention and increase the risks associated with safety-critical work.

Communication

How effective the organisation is at using various communication methods to convey health and safety messages and information out to the workforce, and how well the organisation then checks understanding of those messages is important. Provision of effective methods of communication can encourage workers to behave safely as they will understand the requirements and understand why safety remains important to the organisation. Methods include:

Consultation with workers

- **Levels of supervision** - the presence or absence of, and the competence of, supervision (in the context of health and safety) and the way that poor safety-related behaviour is dealt with. For example, in an organisation that undertakes engineering maintenance work, the presence of competent supervisors to oversee that work is critical as a check to prevent both rule-breaking behaviour and human error.
- **Consultation** and worker involvement - the extent to which workers are involved in the management of health and safety issues and in the decision-making process.
- **Training** - how good the organisation is at identifying health and safety training needs and opportunities and how well it then meets those needs to create well-informed, competent staff.

Job Factors

These are the various characteristics of a worker's job that influence their safety-related behaviour and may involve the following aspects.

Task

This is the characteristics of the work itself, in particular the ergonomic requirements. For example, the need to bend or stoop over when carrying out a task requires the task to be adapted to best suit the worker concerned. In the absence of ergonomic design, workers will find the most comfortable way of working and this may not be the safest way.

> **DEFINITION**
>
> **ERGONOMICS**
>
> The design of the workplace, work equipment and work environment to suit the needs of the individual worker.
>
> Ergonomics is often used as a way of maximising work efficiency, but in the context of health and safety management, it is more concerned with maximising worker comfort and reducing the stress and strain put on the body during work, so as to avoid injuries and ill health (e.g. upper limb disorder caused by repetitive handling activity).

Workload

Workload is the amount of work, rate of work, deadlines and variety of work that individuals have to cope with and the degree to which these are under the direct control of the worker or imposed externally.

Environment

Workplace conditions such as space, lighting, noise, temperature and humidity and the way that these parameters are controlled can minimise their impact on worker performance. For example, workers in a steel foundry may have to undertake physical labour in a high-temperature environment where there is the potential for dehydration, heat stress and heat stroke. Workers may find ways of working that are not necessarily safe, in order to minimise physical exertion. They may also start to suffer degradation in physical and mental performance as a result of heat stress.

Displays and Controls

The design of displays and controls can influence behaviour. Poor designs can contribute to the likelihood of human error such as displays that are difficult to view and critical displays that are out of the operator's normal field of view.

Procedures

The existence of, and quality of, working procedures can affect behaviour. A lack of written procedures or poorly written procedures that are out of date, overly complex or impractical can be a reason for employees not to comply. To be effective, procedures should be accurate, concise, use familiar language and they must be doable.

Individual Factors

People bring to their job their own personal mix of knowledge and experience, skills, attitudes, and personality. These individual characteristics influence behaviour in complex and significant ways. Some of these characteristics cannot be changed, but others can. It is important to recognise where changes to these characteristics might be needed and what methods might best be used to make those changes. For example, if workers have a poor attitude to machine guards, it will be necessary to change their attitude, and there are various ways of attempting to achieve this change. Competence, skills, personality, attitude, risk perception and motivation are individual factors that we will now look at in greater detail.

Competence

Competence is a combination of knowledge, experience, training and ability, that brings a person to a level where they are able to perform to an acceptable standard and where they are aware of their own limitations.

Employers must ensure that workers are competent for the role that they carry out. To be competent, a worker must have the right combination of training and experience - simply having a qualification doesn't necessarily make a person 'competent'. Carrying out the job for a long time doesn't either! A newly qualified person may perform tasks more carefully than a more experienced worker, or be more up to date with current technology, but they may also lack the experience that only comes with time.

Skills

Each individual has skills that they have developed over time; some of these skills are physical (such as the ability of a crane operator to precisely control the movement of a load) whilst others are mental (such as the ability to do complex mental arithmetic whilst working). Fitting a person with the wrong skill set into the wrong role is unwise - a person with poor manual skills may be incapable of safely controlling a piece of machinery during a work activity because they lack the necessary skill.

Personality

These are the various natural characteristics that underpin a person's character, such as risk-taking or being risk averse, introvert or extrovert, etc. We can influence someone's attitude to health and safety, we can increase their skills and develop their competence, but their personality remains largely fixed – it's who we are.

Attitude

Attitude is a person's point of view or way of looking at something, how they think and feel about it.

For example, everyone has an attitude towards work; some people think of it in a positive way and others have a negative attitude. Attitudes develop over time, many of them quite early in life, and they have a tendency to stay with us. This is because they are a part of self-image - the mental picture a person has of themselves and who they are. Attitudes do change but usually only slowly.

In the context of workplace behaviour, attitudes are important because a worker's attitude will make them more or less likely to behave safely. For example, if a worker's attitude to a machine guard is that the guard is great, because it is there to stop their arm being cut off, then that worker is very unlikely to remove the guard in any circumstances.

But if the worker's attitude is that the guard is unnecessary, over-the-top, put there to tick a box or to make the job harder, then that worker is likely to remove the machine guard at the earliest opportunity.

Changing attitudes is notoriously difficult, but can be done using various methods, such as:

- **Education and training** - workers educated in the hazards and risks associated with the machine and the reasons why a guard is necessary may change their attitude over time.
- **High-impact interventions** - workers put through high-impact training, where they are shown pictures of the injuries caused by failure to use the guard, may change their attitude very quickly.
- **Enforcement** - workers forced to use the guard, by supervision and discipline, will find that using it becomes a habit. After a time, attitudes change to match behaviour.
- **Consultation and involvement** in the decision-making process - workers consulted and involved in the selection, design and implementation of the guard will feel more ownership of the process and are more likely to develop a positive attitude towards guard use.

Risk Perception

Perception can be defined as the way that a person interprets information detected by their senses.

Some hazards in the workplace cannot be detected by human senses, (e.g. carbon monoxide gas is colourless, odourless and tasteless yet deadly at relatively low concentrations) so the risk associated with these hazards will not be perceived.

People with some form of sensory impairment may not be able to correctly detect hazards in a workplace. For example, a partially sighted person may not be able to see trip hazards on the floor so they are at greater risk from these hazards than their sighted colleagues.

A worker may not be able to differentiate correctly between red and green and this may create risk to themselves and others. This is why some jobs require an eye test and such difficulties would prevent recruitment (e.g. airline pilots). Both of these examples deal with a sensory impairment (eyesight). Any form of sensory impairment, whether it is sight, hearing, smell, touch or even taste, may mean that a person is unable to correctly perceive the world around them. This might have health and safety implications.

A sensory impairment can be created in the workplace if, for example, conditions prevent the information reaching the senses. Personal Protective Equipment (PPE) is designed to protect but, in many cases, it can cause an impairment (gloves can result in a loss of touch and dexterity, hearing protection prevents warnings being detected, etc.). In addition, background noise, or high levels of odour can also mask issues that we would usually be able to detect.

However, perception goes beyond 'sensory impairment' issues to also cover the way that a person's brain interprets the information sent to it by the senses. A person with fully functioning senses can still make errors in the way that they interpret sensory information.

Optical illusions work using this principle. Look at this picture. Which centre dot is larger?

They are, in fact, the same size. Your eyes work, but the brain interprets the information incorrectly.

A tired worker driving a lorry may not recognise a pothole in the road quickly enough to avoid it, despite the fact that their eyes work perfectly well. It is not their eyes that have failed them, it is the way their brain has interpreted the information sent by their eyes.

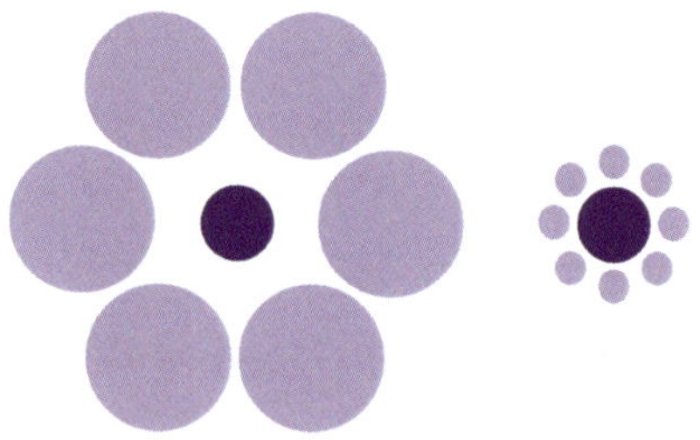

Optical illusion

Factors that can distort a person's perception of hazard and risk include:

- Illness.
- Stress.
- Fatigue.
- Drugs and alcohol.
- Previous experiences.
- Training and education.
- Use of PPE.
- Workplace conditions, such as high levels of noise.

TOPIC FOCUS

Ways of Improving Worker Perception of Hazards

- Carrying out safety awareness campaigns using posters, toolbox talks, etc.
- Developing training programmes to increase awareness of the hazard and its consequences.
- Highlighting hazards, e.g. using safety signs to inform employees that hearing protection is required, or to warn of a hazard, such as the presence of forklift trucks or wet floors. Paint and tape can also be used to highlight hazards like low objects or changes in level (e.g. the edge of steps).
- Ensuring that there is adequate lighting.
- Removing distractions such as noise (which could result in a worker not hearing a warning) or excessive heat (which can cause fatigue).

Motivation

DEFINITION

MOTIVATION

A person's drive towards a goal. The thing that is making them do what they do.

In the context of health and safety, it is important to understand a worker's motivation for carrying out their work in a safe or unsafe way because that motivation can then be changed. Workers often behave unsafely not simply because they are wilful, but because they perceive a reward and they think that the risk is worth the reward. Their unsafe behaviour is incentivised.

For example, a worker who can make more money by taking an unsafe shortcut is far more likely to take that shortcut (if they think they can get away with it). Similarly, a worker who can save themselves time and energy by taking a shortcut is far more likely to do so. If, however, there is no reward, or if the worker thinks that the risk is too great, then they will not take the shortcut.

Any reward scheme in the workplace that is intended to improve safety must be carefully thought through to ensure that it will incentivise the right kind of behaviour. For example, some workplaces pay a bonus related to the number of accidents recorded over a period of time. Fewer accidents equals bigger bonus. This sounds like a good idea because it should make people work more safely so that they have fewer accidents and so get a bigger bonus. But, of course, no one goes to work intending to have an accident. People do not take more care. Instead, what happens is that they report fewer of the accidents that they are having at work so that the number looks better. The incentive scheme actually rewards under-reporting of accidents, not safer behaviour.

MORE...

Further information on human factors can be found at:

www.hse.gov.uk/humanfactors/index.htm

and in the guidance *Reducing error and influencing behaviour* (HSG48) available from the HSE website:

www.hse.gov.uk/pubns/books/hsg48.htm

The Link between Individual, Job and Organisational Factors

The link between individual, job and organisational factors can be seen clearly in the image below.

Accidents can often be prevented by those who have them. If an incident occurs, it is easy to blame the individual worker and to identify the cause as an individual failing. On further investigation, it will probably be seen that an individual's attitude has been formed by a poor organisational culture or there may be work pressures or environmental factors that influenced the person's behaviour.

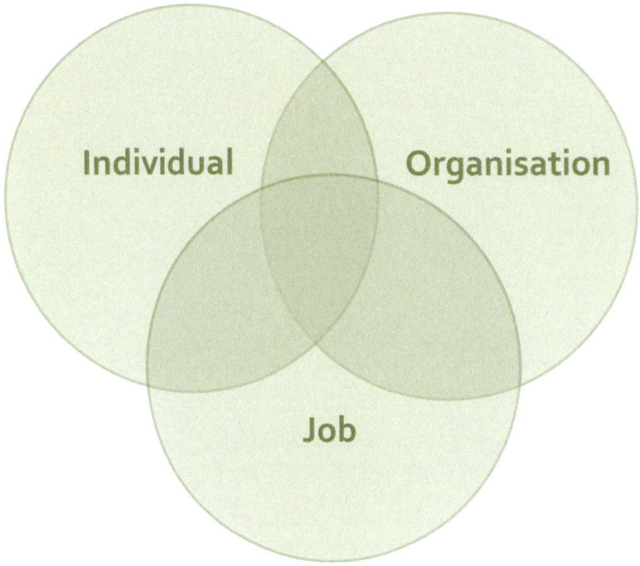

It is only when the three components - job, individual and organisation - work together that a person can be influenced to adopt a safe behaviour

STUDY QUESTIONS

3. Identify the three factors that impact on a person's health-and-safety-related behaviour.
4. Outline the key job factors that can impact on safety-related behaviour.
5. What is risk perception and how may it arise?

(Suggested Answers are at the end.)

Improving Health and Safety Culture

IN THIS SECTION...

- Safety-related behaviour in an organisation can only be improved if a clear commitment has been made by management, with visible leadership.
- Competent staff have the appropriate training, skills, experience and knowledge necessary to do their jobs safely.
- Communication of safety information can be verbal, written or graphic and there are strengths and weaknesses associated with each method.
- Noticeboards, posters, toolbox talks, memos and handbooks all have a part to play in delivering safety information to employees.
- Employers have a legal duty to consult with their workers on health and safety matters.
- This consultation is subject to two sets of regulations: one set applies in the unionised workplace and the other set applies in a non-unionised workplace.
- A health and safety committee is an important method for achieving consultation. To work effectively, the committee has to be set up and run according to agreed rules and procedures, which form part of the policy arrangements of the organisation.
- Training is a vital tool in improving safety-related behaviour and there are various occasions when training should be provided, for example, at induction.

Gaining Management Commitment

Lack of management commitment will suggest that health and safety is not a high priority.

Gaining senior management commitment, therefore, is a vital step when trying to move towards a more positive health and safety culture. Management commitment starts at the very top of the organisation. Senior managers must provide the leadership necessary to inspire and motivate managers at all levels to pursue health and safety objectives rigorously. This is done by establishing the organisation's health and safety policy with clear priorities and targets to be achieved.

It is also critical that middle and junior management follow through the commitment of senior management by means of the priorities and objectives that they set their staff. In this way, commitment is cascaded down through the organisation.

Management commitment and leadership

Promoting Health and Safety Standards by Leadership, Example and Disciplinary Procedures

An important factor in demonstrating management commitment is visible leadership. If management is never seen taking an active interest in health and safety issues, then there will be an assumption that they are not interested. Individual managers must show their commitment to health and safety to their staff as this creates the local safety culture.

Visible leadership can be demonstrated by:

- Behaving safely (leading by example).
- Involvement in the day-to-day management of health and safety, e.g. by attending safety meetings.
- Taking part in safety tours or audits.
- Promoting changes to improve health and safety.
- Involvement in investigation of incidents and ill health.
- Enforcing the company rules

On the last point, managers must be able to recognise and enforce the correct health and safety standards. They must be able to praise and reward good behaviour and they must be able to challenge and correct poor health- and safety-related behaviour. Workers must not be allowed to carry out their work unsafely without being challenged. If this is allowed to persist, management will quickly lose the ability to challenge any unsafe behaviour and will be unable to enforce discipline. This means that sometimes workers have to be taken into formal disciplinary procedures for failure to comply with health and safety requirements. Their behaviour may be so bad that this results in dismissal for gross misconduct.

Competent Workers

A competent person is someone who has sufficient training, skills, experience and knowledge (and perhaps other abilities such as attitude and physical ability) to be able to carry out their work safely.

It is the responsibility of the employer to ensure that workers are competent to carry out the tasks that they have been allocated. The more competent the worker, the better able they will be to do their job safely. This has a positive influence on health and safety culture. In order to determine competence, the employer may check qualifications, request references, or verify membership of professional bodies.

Managers should also be competent. This means that all managers should have an understanding of the health and safety implications of the decisions they make on a day-to-day basis. This is often overlooked. For example, if a manager is in control of a construction site then they must understand the difference between safe and unsafe scaffolding. They do not need to be able to erect the scaffold themselves, but they must have sufficient knowledge to spot good and poor practice when they see it.

Good Communication within the Organisation

Communication can be defined as the process of delivering information from a sender to a recipient. To be truly effective, the correct information has to be transmitted, received and understood.

There are three principal delivery methods for communicating information: **verbal**, **written** and **graphic**.

Benefits and Limitations of Different Methods of Communication

Verbal Communication

Communication using the spoken word, e.g. face-to-face conversations, meetings, interviews, training sessions, by telephone or over a Public Announcement (PA) system.

Verbal communication at work

This is the easiest and most commonly used form of communication, but there are various weaknesses associated with this method. If verbal communication is to be used to convey safety-critical information to workers, these weaknesses must be overcome.

<table>
<tr><th>Merits</th><th>Limitations</th></tr>
<tr><td><ul><li>Personal.</li><li>Quick.</li><li>Direct.</li><li>Allows for checking of understanding.</li><li>Allows for feedback to be given.</li><li>Allows for exchange of views.</li><li>Usually allows for additional information to be transmitted by means of tone of voice, facial expression and body language.</li></ul></td><td><ul><li>Language barrier may exist.</li><li>Jargon may not be understood.</li><li>Strong accent or dialect may interfere.</li><li>Background noise may interfere.</li><li>Recipient may have poor hearing.</li><li>Message may be ambiguous.</li><li>Recipient may miss information.</li><li>Recipient may forget information.</li><li>No written record as proof.</li><li>Poor transmission quality if by telephone or PA system.</li></ul></td></tr>
</table>

Written Communication

Communication using the written word, e.g. report, memo, e-mail, notice, company handbook, policy document, operating instructions, risk assessment, minutes of meetings, etc.

<table>
<tr><th>Merits</th><th>Limitations</th></tr>
<tr><td><ul><li>Permanent record.</li><li>Can be referred back to.</li><li>Can be written very carefully to avoid use of jargon, abbreviations and ambiguity.</li><li>Can be distributed to a wide audience relatively cheaply.</li></ul></td><td><ul><li>Indirect.</li><li>Takes time to write.</li><li>May contain jargon and abbreviations.</li><li>Can be impersonal.</li><li>Message may be ambiguous.</li><li>Message may not be read by recipient.</li><li>Language barrier may exist.</li><li>Recipient may not be able to read.</li><li>Immediate feedback is not available.</li><li>Questions cannot be asked.</li><li>Recipient may have impaired vision.</li></ul></td></tr>
</table>

Graphic Communication

Communication using pictures, symbols or pictograms, e.g. safety signs, such as a fire exit sign; hazard-warning symbols, such as a skull and crossbones found on the label of a toxic chemical; or photographs, such as of a machine showing a guard being used correctly in the operating instructions for the machine.

Merits	Limitations
• Eye-catching. • Visual. • Quick to interpret. • No language barrier. • Jargon-free. • Conveys a message to a wide audience.	• Can only convey simple messages. • Might be expensive to buy or produce. • May not be looked at. • Symbols or pictograms may be unknown to the recipient. • No immediate feedback available. • Questions cannot be asked. • Recipient may have impaired vision.

Graphic communication

Use and Effectiveness of Noticeboards and Health and Safety Media

There are various ways of broadcasting health and safety information using the three methods we discussed above. Each of these broadcasting techniques has its own strengths and limitations, and so usually a mix of some or all of these techniques is used to ensure that essential messages are transmitted and correctly understood by all staff.

- **Noticeboards** - should be eye-catching and located in areas used by all workers, e.g. rest rooms or central corridors. Notices should be current, relevant and tidily displayed. Cluttered, out-of-date, irrelevant notices obscure the messages being conveyed. Displaying a notice does not mean that it will be read. Typical contents might include: the safety policy; employer's liability insurance certificate; emergency procedures; identity of safety representatives and first aiders; minutes of safety committee meeting; accident statistics, etc.
- **Posters** - used to provide safety information, drawing attention to particular issues and supporting the safety culture.

Merits	Limitations
• Graphic and therefore avoid language barriers. • Can be eye-catching and generate interest. • Low cost. • Can reinforce key messages.	• Can quickly become part of the surroundings. • May be defaced. • Can trivialise important issues. • Rely on the recipient interpreting the correct message from the image.

One poster that must be displayed in the workplace is the 'health and safety law' poster. The example shows the poster displayed at RRC's Head Office.

Example of the health and safety law poster

It is a legal requirement under the **Health and Safety Information for Employees Regulations 1989** that this poster is displayed in a central location within the workplace where employees can read it. The poster has blank spaces for the details of worker representatives and other useful contacts. If workers do not work from a central location, (e.g. home workers), then a leaflet can be provided that contains the same information.

- **Films or videos** - mainly used in training programmes and, if well made, can hold the audience's attention. It is a relatively cost effective medium, and a large audience can be reached in a short time. Important points can be made strongly and vividly.

  However, it is difficult to get feedback regarding their effectiveness, they can very quickly become out of date and irrelevant and it might tempt some employers to think that this is all they need to do.

- **Digital media and Intranet systems** - many organisations publish the majority of their health and safety management information, instructions and policy arrangements in electronic form. In many cases, key documents are only available from the organisation's Intranet system (the internal IT network). It is almost entirely reliant on workers willingness to access the system and, on some occasions, the technology may not be available.
- **Toolbox talks** - short, practical safety briefings carried out routinely in the workplace, often presented by the supervisor at the start of a shift. They can be useful for generating awareness and discussion on safety precautions, but seen as dull or a waste of time if topics are irrelevant or poorly presented.
- **Memos and e-mails** - written notifications used to provide specific information about a single issue, such as updating procedures, drawing attention to lapses in practice, etc. A wide audience can be reached quickly and cost effectively and there is a written record of the communication. There is often limited opportunity for feedback or questioning, so their use is really limited to issuing clear and precise instructions or information.
- **Employee handbooks** - used to set out the organisation's health and safety policy. All employees should be given a copy on joining the organisation and updates are usually circulated to inform staff of changes. This is a key document, containing information such as site rules, reporting procedures, emergency arrangements, etc. It is standard practice to issue this handbook as part of the worker's induction and to get a receipt as proof of issue.

Key health and safety management information, such as procedures and instructions, must be broadcast using a range of methods to ensure it reaches the correct audience and to ensure that the various barriers to communication associated with different communication methods are overcome.

The goal is to ensure that the correct message has reached the right people and has been properly understood.

Co-operation and Consultation with the Workforce and Contractors

Improvements to health- and safety-related behaviour will only occur in an organisation with worker co-operation and involvement. If workers feel that they are being dictated to then they will feel little ownership of health and safety. Indeed, they may come to resent instructions being imposed from on high and start to actively oppose health and safety initiatives and improvements. This creates a negative culture and negative behaviour.

The most effective way to avoid this negativity and to actively encourage worker interest and ownership is to involve workers in the decision-making process. This is best done through the consultation process.

DEFINITION

CONSULTING

The two-way exchange of information and opinion between the employer and workers so that the best course of action can be agreed. This implies that the employer listens to the concerns of his workers and changes his plans as necessary.

Consultation implies that the employer takes note of feedback given by employees. But the consultation process is not the only route by which feedback on health and safety topics might be given. Employee feedback can also be gained through additional routes such as during appraisals, at departmental meetings, through suggestion schemes, through employee surveys and via complaints.

The consultation process will always require an employer to consult with their own employees. Sometimes, it will also require consultation with other workers, such as contractors working within the employer's premises or undertaking work on behalf of the employer. The following section focuses on the legal requirements for an employer to consult with their employees. It is worth mentioning that, alongside the legal requirement highlighted below, there may be a legal requirement to consult with contractors, e.g. **CDM Regulations 2015** (see Element 1).

The Appointment, Functions and Entitlements of Worker Representatives

There is a legal duty placed upon employers to consult with their employees on health and safety matters. This legal duty is defined in two sets of regulations:

- The **Safety Representatives and Safety Committees Regulations 1977** which cover unionised workplaces.
- The **Health and Safety (Consultation with Employees) Regulations 1996** which cover non-unionised workplaces.

These two sets of regulations are broadly similar, though there are differences. The diagram that follows shows the rough outline of the consultation framework that exists under these two sets of regulations.

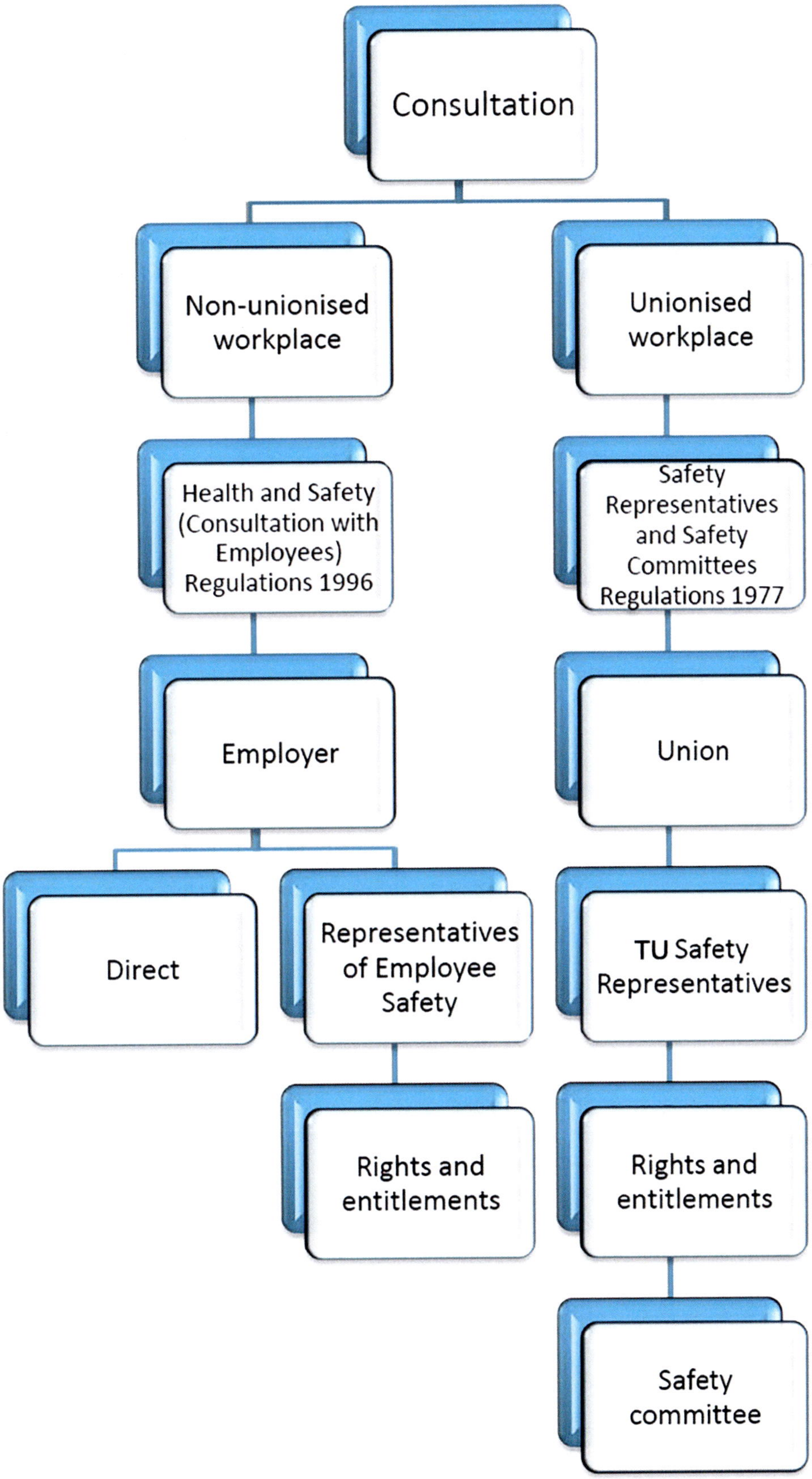

Consultation framework

Under the two sets of regulations, an employer does not have to consult with employees on everything, but particular health and safety issues where consultation would be appropriate include:

- The introduction of measures affecting the health and safety of the workers.
- The appointment of safety advisers and specialists.
- Health and safety training plans.
- The introduction of new technology into the workplace that will affect health and safety.

Trade Union Safety Representatives

Under the **Safety Representatives and Safety Committees Regulations 1977**, a recognised, independent trade union has the right to appoint safety representatives. These union members will ideally have two years' relevant experience. Once the union notifies the employer of the appointment of the safety representatives in writing, then the employer must consult with them on health and safety matters relating to the groups of workers that they are representing (in effect, their constituency).

TOPIC FOCUS

Safety representatives have a number of rights and entitlements, such as the right to:

- Carry out an inspection of the workplace (at least once every three months).
- Examine the causes of accidents.
- Examine and copy records and documents relating to health and safety.
- Receive information from Health and Safety Executive (HSE) inspectors.
- Investigate complaints about health and safety.
- Make representations to the employer.
- Be consulted on health and safety matters.
- Time off with pay to perform their functions.
- Time off with pay for appropriate training.
- Reasonable facilities to perform their functions.

If the employer does not allow the safety representative to carry out these rights and entitlements, then the employee can appeal to an employment tribunal. Safety representatives are also given protection from being treated detrimentally by other employment law. They do not have any criminal or civil liability for health and safety standards in the workplace; for this reason they have 'rights and entitlements' rather than 'duties'.

Representatives of Employee Safety

Employee safety Under the **Health and Safety (Consultation with Employees) Regulations 1996** where the employer does not recognise a trade union for bargaining purposes, they can decide to consult with their employees directly or through elected representatives. If the employer chooses the latter route then these representatives are called Representatives of Employee Safety (RoES).

Employee safety consultation

Should an employer not wish to have RoES, they are allowed to consult directly with their employees. This could be done at briefings, toolbox talks, departmental meetings, appraisals, etc. Many small organisations with a few tens of employees might consult by this direct method because it is quickly and easily achieved. However, in organisations of any significant size, this direct consultation model becomes difficult, inefficient and ineffective. It makes more sense to have elected RoES and meeting with these representatives is the usual method of consultation.

Once elected, RoES have very similar functions, rights and entitlements to their union equivalents with some notable exceptions: they are not entitled to carry out an inspection of the workplace, or investigate the causes of accidents, receive certain information from inspectors or attend a safety committee. Of course, this does not prevent them from doing so with the employer's permission, but they are not legally entitled to do so.

Safety Representatives and Safety Committees Regulations 1977	**Health and Safety (Consultation with Employees) Regulations 1996**
Representatives:	
Appointed in writing by a trade union recognised for collective bargaining purposes.	Elected by the workforce, where the employer has decided not to consult directly.
Title/position:	
Safety representatives.	Representatives of employee safety.
Functions:	
Investigate potential hazards and dangerous occurrences at the workplace, complaints by an employee relating to health, safety and welfare at work, and examine causes of workplace accidents.	
Representation to the employer on the above investigations, and on general matters affecting the health and safety of the employees they represent.	Representation to the employer on: • potential hazards and dangerous occurrences; • general matters affecting the health and safety of the employees they represent; and • specific matters on which the employer must consult.
Inspect the workplace.	
Represent employees in dealings with health and safety inspectors.	Represent employees in dealings with health and safety inspectors.
Receive certain information from inspectors.	
Attend health and safety committee meetings.	

Source: INDG232(rev2) *Consulting employees on health and safety*, HSE, 2013 (www.hse. gov.uk/pubns/indg232.pdf)

Safety Committees

Under the **Safety Representatives and Safety Committees Regulations 1977**, an employer is obliged to establish a safety committee when requested to do so by two or more safety representatives in writing. The employer must comply within three months.

An employer is not under a similar obligation under the **Health and Safety (Consultation with Employees) Regulations 1996**. However, since an employer is obliged to consult with RoES under these regulations, the formation of a safety group or safety team is the usual route taken.

HINTS AND TIPS

You will note for non-unionised organisations, we have used the term 'safety team' rather than 'committee'. In practice, it can be called anything you like, but for the exam it is a good idea to use the terms 'reps and committees' for unions and 'RoES and team' for non-unions.

For the rest of this section, references to the effective safety committee apply equally to any safety group.

The exact composition of the safety committee is a matter for agreement between the employer and the representatives. Members might include:

- Worker safety representatives.
- One or more operational managers with authority.
- A safety adviser.
- An occupational health nurse or doctor.
- Facilities manager.
- Human resources manager.
- Training manager.

TOPIC FOCUS

The Health and Safety Committee

This applies equally to any health and safety team or group.

To work effectively, the committee has to be set up and run according to agreed rules and procedures, which form a part of the policy arrangements of the organisation. The following issues should be taken into account in these arrangements:

- Who is on the committee? There has to be a balance between managers and workers, and the right managers have to be included.
- How often will the committee meet? The committee should meet regularly and frequently enough to be useful, e.g. once a month.
- Who will act as chairperson? All meetings need someone to take charge so that the discussion during the meeting is relevant, and to make sure that everybody has an opportunity to speak.
- What authority will the committee have? The committee must be able to make decisions, otherwise there will be lots of talk but no action. Usually, the committee will involve one or more senior managers who have executive authority.

(Continued)

TOPIC FOCUS

What will be discussed? It is common practice for a committee meeting to have a published agenda that has been agreed before the meeting takes place.

- How will the discussions be recorded? Minutes of the meetings are usually taken and then circulated to all attendees and posted on noticeboards in the workplace for all workers to see.
- How will issues discussed be followed up? All agreed actions must be recorded in the minutes of the meeting together with the name of the person responsible for taking that action and a deadline. This **action plan** can then be reviewed at the start of the next meeting to check that the action has been completed.

The exact functions of the safety committee are not dictated, but should include issues such as:

- Studying accident and disease statistics.
- Reviewing the reports from active monitoring in the workplace, such as safety inspections and behavioural observations.
- Examining safety audit reports.
- Considering reports and information from HSE.
- Considering reports submitted by safety representatives.
- Assisting in the development of procedures and policy.
- Monitoring the effectiveness of training.
- Monitoring the effectiveness of safety communications.

In this way, the committee is intended to be strategic, rather than involved in trivia.

MORE...

Further information on leadership and consultation can be found at:

www.hse.gov.uk/pubns/indg277.pdf

www.hse.gov.uk/involvement/index.htm

Benefits of Worker Participation

There are many benefits of employee participation. They include:

- The development of a stronger employee team. Employees that feel their opinion is valued and see changes designed to benefit both them and their colleagues and are more likely to feel invested in the success of the company.
- It breaks down traditional communication barriers. Employees that have a forum for speaking their thoughts and fears and are more likely to point out problems or better ways of doing things.
- Improving participation helps explain and build acceptance of new changes.
- Participation can also serve to reduce stress in both employees and employers. Stress can build up over uncertainties regarding company changes, because of methods of work, or because employee ideas are not taken into consideration.

Worker participation is seen as one of the biggest influences on health and safety behaviour and culture.

When Training is Needed

Training (in the context of health and safety) can be defined as the planned, formal process of acquiring and practising knowledge and skills in a relatively safe environment.

Training is central to the management of health and safety in workplaces. Employers have a responsibility to train their staff to carry out their jobs in a safe manner. Training is a key component of competence. In the absence of training, it is difficult to develop or demonstrate competence and, as a result, statute law in many countries requires that an employer provides appropriate training for their workers.

The reason for this requirement is simple; training has a dramatic effect on health and safety-related behaviour. Without training, workers try to do their jobs to the best of their ability but they do so either by informally copying others (including copying all the bad habits and unsafe working practices that they see) or by doing the job the way that they think is best. Once the worker has been trained they will understand:

- The hazards and risks inherent in their work.
- The correct rules and precautions to apply.
- Foreseeable emergencies and the actions to take should these events occur.
- Identifying who to contact with any issues.
- Limitations and restrictions that apply to their work.
- Their personal health and safety responsibilities.
- The consequences of breaking the rules, including disciplinary procedures.

Various circumstances, such as the ones described below, require the provision of training.

Induction

Induction training takes place when workers join an organisation. This allows the worker to obtain knowledge about the organisation in a safe, structured manner and ensures that critical information is delivered and understood. Since a worker is at risk in a workplace from their first day of work, it makes sense to deliver induction training as soon as they start work and to cover safety-critical information, such as emergency procedures first.

TOPIC FOCUS

Typical content of a **general induction training course** for new starters includes:

- The organisation's health and safety policy.
- Fire and other emergency procedures.
- Details of specific workplace hazards and controls.
- First-aid facilities and personnel.
- The location of welfare facilities.
- Safe movement around the workplace.
- Accident- and incident-reporting procedures.
- Worker consultation arrangements.
- General safety rules, such as no smoking areas.
- PPE requirements.
- Introduction to the Safe Systems of Work (SSOWs) and permit systems.
- Introduction to the risk assessment system.
- Responsibilities of individuals.
- Disciplinary procedures.

Job Change

Additional training is necessary when a worker's job changes so that they are exposed to new hazards and risks. For example, a construction worker whose job changes from working on a construction site to being a service engineer will need additional training, not in how to service equipment - since the type of work has not changed - but in lone working. They will be at far greater risk when they go out into the community to conduct home visits compared with working on a construction site.

Process Change

When the way in which the work is done changes, workers may be exposed to new hazards and risks that require additional training. For example, when a different product is being produced on an existing piece of machinery, this may create new risks that require training in new safe operating procedures.

Introduction of New Legislation

Changes to the law governing a particular health and safety issue often create a need to train workers on the implications of the new legislation, perhaps because working practices have to change or simply to ensure an understanding of the law and its requirements.

Introduction of New Technology

New technologies adopted by organisations create different hazards and risks that workers may be unfamiliar with. The mass introduction of desktop computers, screens and keyboards in the 1990s is an example of new technology introducing new risk into workplaces. This has been further compounded with the widespread introduction and use of mobile phones and tablets which bring additional risks and complications to the safe use of IT.

It is important to keep records of any training given, not only to record who has been trained, but also to update and show progress against a training plan. Training records should detail the level of competence achieved, the date of the training and highlight when any refresher training is required. These records may be used to demonstrate to the authorities that adequate training was provided, or to prove that an employee had received training in the event of a civil claim for compensation or an accident investigation.

STUDY QUESTIONS

6. What are the main merits (advantages) and limitations (disadvantages) of both written and verbal forms of communication?
7. How are graphic symbols (pictograms) used in safety communications?
8. What should be the first priority in induction training?
9. Apart from at induction, when should training be provided?
10. How can employees be involved in the improvement of workplace health and safety?
11. How must employers consult employees?
12. What rights do trade union safety representatives have?
13. When must an employer establish a safety committee?

(Suggested Answers are at the end.)

Assessing Risk

IN THIS SECTION...

- Risk assessment is the formalised process of identifying hazards, evaluating risk and then either eliminating or controlling that risk to an acceptable level and is a legal requirement under the **Management of Health and Safety at Work Regulations 1999 (MHSWR)**.
- A hazard is something with the potential to cause harm.
- Risk is the likelihood that a hazard will cause harm in combination with the severity of injury, damage or loss that might foreseeably occur.
- The main objective of risk assessment is the prevention of accidents and ill health.
- There are five steps to risk assessment:
    1. Identify the hazards.
    2. Identify the people who might be harmed and how.
    3. Evaluate the risk and decide on precautions.
    4. Record the significant findings and implement them.
    5. Review and update as necessary.
- Employees, contractors, visitors and members of the public must all be considered in the risk assessment process.
- If the risk is unacceptable, then controls must be introduced based on the general hierarchy of control.
- Sometimes it is necessary to focus risk assessment on a vulnerable person or group of workers such as young persons, nursing and expectant mothers, disabled workers and lone workers.

Meaning of Hazard, Risk and Risk Assessment

Hazard

DEFINITION

HAZARD

Something with the potential to cause harm.

This can include articles, substances, plant or machines, methods of work, the working environment and other aspects of work organisation.

Hazards can be broadly classified as:

- **Physical** - things that cause harm because of their physical characteristics, e.g. electricity, work at height, radiation, vibration, noise, heat, trip hazards, moving machine parts, vehicles, etc.
- **Chemical** - things that cause harm because of their chemical characteristics, e.g. lead, mercury, sulphuric acid, silica, cement dust, etc.

A wet floor is a hazard that creates the risk of slips

- **Biological** - living microorganisms that cause disease and ill health, e.g. hepatitis B virus (HBV), *Legionella* bacteria (responsible for Legionnaires' disease), Rabies virus, etc.
- **Ergonomic** - stress and strain put on the body through posture and movement, e.g. frequent repetitive handling of small boxes leading to inflammation of the tendons in the elbow joint.
- **Psychological** - things that have the potential to cause injury to the mind rather than the body, e.g. exposure to highly traumatic events that can leave a person unable to adjust to a normal life after the event (a condition referred to as Post-Traumatic Stress Disorder (PTSD)).

For example, a lorry moving around a site road is a physical hazard because it might run over a worker. Sodium hydroxide (caustic soda) is a chemical hazard because it is a highly alkaline chemical capable of causing corrosive burns.

Note that a hazard is the 'something' that causes the harm. If an office worker receives an electric shock from an item of electrical equipment that has a damaged flex, then electricity is the hazard, not the damaged flex. It is electricity that causes the harm; the damaged flex is the failure in the controls or preventive measures. If the flex were not damaged then the hazard would still be present (electricity is still running through the equipment) but it would be properly controlled and the electric shock would not occur.

Risk

DEFINITION

RISK

The likelihood of potential harm from that hazard being realised, in combination with the severity of injury, damage or loss that might foreseeably occur.

Risk can be described **qualitatively** using words such as 'high', 'medium' or 'low'. There will always be a degree of subjectivity to this qualitative description since the words represent one person's opinion of the risk level. Different individuals have very different personality characteristics and so two people may disagree on the level of risk inherent in a hazard.

Risk can also be defined **quantitatively** using probabilities and/or frequencies that have been derived from hard data. This type of quantified risk assessment is far more rigorous than qualitative risk assessment but is beyond the scope of this course.

Risk Assessment

DEFINITION

RISK ASSESSMENT

Risk assessment is the formal process of identifying preventive and protective measures by evaluating the risks arising from a hazard, taking into account the adequacy of any existing controls, and deciding whether or not the risk is acceptable.

Risk assessment is a process that people do automatically all the time. When you cross the road you carry out a risk assessment; when you drive a car you carry out a risk assessment; when you boil a kettle you carry out a risk assessment. But, of course, this assessment is normally done very quickly and without conscious thought or effort. If you are not very good at this process then you will not live long.

There are occasions in normal life, however, when you might become more aware that you are assessing risks. If you look after very young children you will consciously think about the particular hazards that present a risk to a child. If you start to take part in certain sports or activities, such as rock climbing or scuba diving, you will start to assess risks in your conscious mind rather than doing it automatically.

A workplace risk assessment is simply an extension of this automatic self-preservation mechanism that has been formalised to meet legal requirements.

Purpose of Risk Assessment and the 'Suitable and Sufficient' Standard it Needs to Reach

Purpose

The purpose of risk assessment is to ensure that hazards are eliminated or risks minimised by the correct application of relevant standards, to prevent workplace accidents.

In particular, the objectives of risk assessment are to prevent:

- Death and personal injury.
- Ill health.
- Other types of loss incidents.
- Breaches of statute law which might lead to enforcement action and/or prosecution.
- The direct and indirect costs that follow on from accidents.

These objectives relate directly to the moral and financial arguments we discussed in Element 1.

Suitable and Sufficient Standard

A risk assessment should be 'suitable and sufficient'.

In other words, it should be good enough to fulfil legal requirements (**MHSWR**) and prevent foreseeable injuries and ill health from happening.

TOPIC FOCUS

A **suitable and sufficient** risk assessment should:

- Be carried out by competent assessors
- Identify the significant risks arising out of work, i.e. those that are most likely to occur and result in harm being caused, with any remaining risks being at an acceptable low level.
- Identify all those persons who could be at risk, including workers and others such as visitors. Vulnerable people, such as young persons, should also be identified.
- Evaluate the effectiveness of any current controls.
- Enable the employer to identify and prioritise the measures that must be taken to protect people from harm, including complying with any relevant legal provisions.
- Be appropriate to the nature of the work and remain valid for a reasonable period of time. In other words, the assessment should be proportionate to the risks in the workplace.

To expand on this last point, that a risk assessment should be proportionate to the risks inherent in the workplace:

- A low-risk workplace with a few straightforward, often predictable hazards, (e.g. a retail shop), should have a relatively simple risk assessment carried out by a competent person (perhaps the manager) by reference to some basic guidance documents.
- A high-risk workplace, (e.g. a chemical works), should have a far more complex risk assessment carried out by competent persons (PhD industrial chemists, etc.) using detailed, complex reference material.

The first assessment might take a few hours to complete; the second might take weeks.

Risk assessments should be carried out by competent people. In this context, the word 'competent' would mean people who have sufficient training, knowledge, experience and other abilities. The exact training, knowledge and experience required will vary depending on circumstances. In some instances, simply an ability to identify, read and correctly interpret guidance on a topic is sufficient. In others, a detailed understanding of background knowledge is essential to be able to correctly evaluate risk.

Workers familiar with the tasks and areas to be assessed

A risk assessment might be carried out by one person. This is not ideal in many instances since it relies on one person's opinion and judgment. Ideally, risk assessment will be carried out by a team. This allows for various views and opinions to be taken into account and so may result in a more successful assessment. The composition of a risk assessment team is not dictated, but might include:

- Workers familiar with the tasks and areas to be assessed.
- Health and safety specialists such as safety practitioners and occupational health nurses.
- Technical specialists such as mechanical and electrical engineers.
- Line managers responsible for the tasks or areas being assessed.
- Worker safety representatives.

The size and composition of the team will vary depending on the nature of the workplace and the complexity of the risk assessment process being used.

Note that it is not necessary for all members of the team to be competent in the risk assessment process, simply for some or one of the team members to be a competent person.

The involvement of non-competent persons (those who aren't trained risk assessors) is useful for a number of reasons:

- Those team members may identify hazards and risks that might otherwise be missed (two pairs of eyes are better than one).
- They may ask questions and propose solutions that might otherwise not be considered.
- It allows experience to be safely gained in the practice of risk assessment.
- It facilitates employee awareness, involvement and consultation and hence enhances safety culture.

A General Approach to Risk Assessment

The HSE has defined the following steps to risk assessment:

1. Identify the hazards.
2. Identify the people who might be harmed and how.
3. Evaluate the risk and decide on precautions.
4. Record the significant findings and implement them.
5. Review and update as necessary.

Step 1 - Identify the Hazards

The first step in the risk assessment process is to identify all the significant hazards associated with the work.

As discussed earlier, hazards are the things with the potential to cause harm. It is important to identify both the **safety hazards** that might give rise to immediate physical injury and the **health hazards** that might cause disease or ill health.

Safety Hazards

There are many hazards that are capable of causing immediate physical injury:

- Work at height.
- Stacked materials.
- Moving vehicles.
- Manual handling operations.
- Moving parts of machinery.
- Electricity.
- Toxic or corrosive chemicals.
- Deep water.
- A naked flame, e.g. lit propane torch.
- Animals such as dogs.
- Violent people.

Health Hazards

Some hazards can cause occupational disease or ill-health conditions. This can follow a single event (such as a needlestick injury infecting a worker with the hepatitis virus), but more often occurs as a result of prolonged exposure to the hazard over a period of weeks, months or years. These hazards might be referred to as **health hazards** and can be categorised into five groups:

- Physical, e.g. radiation, vibration, noise, extremes of temperature, etc.
- Chemical, e.g. lead, mercury, sulphuric acid, silica, cement dust, etc.
- Biological, e.g. hepatitis B virus (HBV), *Legionella* bacteria (responsible for Legionnaires' disease), Rabies virus, etc.
- Ergonomic, e.g. very repetitive movement, stooping, twisting, manual handling, etc.
- Psychological, e.g. stress and trauma.

Exposure to high noise levels can cause noise-induced hearing loss

Hazard Identification Methods

There are various methods that can be used in hazard identification, such as task analysis, reference to legal guidance, manufacturers' information and incident data.

Task Analysis

This is a useful method for identifying hazards, since it allows hazards to be spotted before work starts, rather than after the work has commenced. Task analysis involves breaking a job down into component steps and identifying the hazards associated with each step, so that the safe working method can then be established to deal with each hazard. This can be done before work starts as part of the planning process, and is how SSOWs are developed.

TOPIC FOCUS

There is a useful acronym for task analysis, **SREDIM**:

- **S**elect the task to be analysed.
- **R**ecord the steps or stages of the task.
- **E**valuate the risks associated with each step.
- **D**evelop a safe working method.
- **I**mplement the safe working method.
- **M**onitor to ensure it is effective.

Legislation

Knowledge of the legal standards that apply to a particular workplace is an important aid when identifying significant hazards. For example, knowledge of the law relating to work at height will allow a competent assessor to identify which work might fall within the definition of work at height and which work could be ignored. Legislation is often accompanied by guidance documents which can be very useful in the identification of hazards. For example, guidance documents exist to spell out all the hazards that exist in engineering workshops.

Manufacturers' Information

When a new item of plant, machinery or equipment is purchased, it usually comes with an instruction book that contains information about all the related hazards and instructions for safe use, cleaning and maintenance. Similarly, when a new substance is purchased, it comes with labels and a Safety Data Sheet (SDS) that clearly identify the hazards of the substance.

Incident Data

Internal accident and near-miss data can be useful in identifying hazards. The main limitation here is that a hazard may be very significant but may not yet have caused harm in the organisation and may therefore go unnoticed. External data, such as national statistics published by the authorities, can be more useful since it identifies the real hazards and risks based on a much larger population size.

Guidance

The HSE produces a wide range of guidance documents and Approved Codes of Practice (ACoPs). All current legislation is accompanied by one or both of these publications. They relate to all industries and cover a wide range of issues. The subject matter includes, for example, confined spaces, working at height, substances and radiation, etc.

Remember that a risk assessment is a tool for identifying all the significant hazards that exist in a workplace - all the things that have the potential to cause harm. It is not a tool for only identifying those hazards that are poorly controlled. So, for example, in a new office with modern computer screens and keyboards, where someone has left a pile of boxes in front of a fire exit door, the hazards are electricity, use of Display Screen Equipment (DSE), fire and poor housekeeping. Not just the pile of boxes in front of the fire exit door, because that would ignore all the other hazards that exist in the office. In an office, you are arguably at far greater risk from death by electric shock than death by boxes left in front of a fire exit door. The first assessment takes the electrical hazard into account, the second ignores it and therefore fails.

Step 2 - Identify People at Risk

When identifying people at risk, think not only of those carrying out particular activities, but also of those who may be affected by those activities. Individuals do not need to be named; rather, general groups or populations identified.

Workers and Operators

Everyone on a construction site can be regarded as workers. For example, an operator is a worker whose role is to use and control vehicles or plant. They may be directly involved with the activity by erecting barriers (worker) or using a mechanical digger (operator). They may be working or passing nearby. Some hazards create risk only for the worker carrying out the task (e.g. a worker up a ladder is at risk from falling) while others create general risk for all workers (e.g. a vehicle traffic route that all workers may have to cross in order to reach a staff car park)..

Maintenance Staff

They are often involved in the removal of the usual safeguards present in the workplace because of the nature of maintenance work, (e.g. the lift engineer who has to climb onto the top of a lift carriage in the lift shaft, or the engineer who has to remove machine guards to repair a breakdown). If the normal safeguards are being removed or bypassed, then risk to these workers increases and other methods have to be found to control this risk.

Cleaners

Cleaners may be exposed to greater risk because cleaning work may involve the removal of safeguards or activities that create additional risk (e.g. window cleaning from an access cradle). Many cleaners also work alone, outside normal working hours, and therefore lone working also becomes an issue.

Contractors

Contractors may be carrying out work independently of the work being carried out by employees or may be working alongside employees. The workplace creates risks for these contractors and the contractors create risks for the workplace. All these risks have to be considered through the risk assessment process.

Visitors

Visitors may not be working but are still exposed to certain types of risk (e.g. fire). Visitors' perceptions of workplace hazards may be different from that of workers' and they may behave differently due to their lack of awareness.

Members of the Public

They may simply be in the vicinity of the workplace, yet are still affected by certain types of hazard. For example, a release of toxic chlorine gas from an industrial site will affect passers-by and those who live near the site. In some instances, trespassers (uninvited visitors) may get onto the site. This is particularly important with regard to the possibility of children coming onto the premises, (e.g. playing on building sites or near railway lines).

In certain instances, identifying general groups of people who might be harmed by hazards is inadequate and a more specific focus has to be applied to a particular person or type of person who is more vulnerable for one reason or another. Young people, nursing and expectant mothers, disabled workers and lone workers all present these special cases (see later in this element).

Step 3 - Evaluate Risk (Taking into Account What You Already Do) and Decide on if You Need to Do More

Having identified a particular hazard and the people who might be harmed by it, the next step in the risk assessment process is to answer a simple question:

Is the level of risk generated by the hazard acceptable or does it need to be reduced?

The question may be simple, but the answer can at times be complex.

Likelihood of Harm and Probable Severity

As previously discussed, risk is a combination of the likelihood that a hazard will cause harm with the foreseeable severity of the injury, should harm occur.

Risk can be qualitatively described using words such as 'very high', 'high', 'medium', 'low' or 'insignificant'. The problem with these (or similar) words is that they are opinion-based and it may therefore be difficult to achieve consistent outcomes from their use.

An alternative approach that is commonly adopted is to break risk down into its two component parts and define each separately:

Risk = Likelihood × Severity

Risk Rating

By the simple substitution of a word for a score, it is possible to calculate a risk rating for a particular hazard. For example:

Likelihood	Severity
1 = extremely unlikely	1 = very minor injury
2 = unlikely	2 = first-aid injury
3 = possible	3 = lost time injury
4 = likely	4 = hospital treatment
5 = very probable	5 = disabling injury

Using this scoring system, the risk generated by a trailing electrical flex positioned across a busy corridor might be calculated as 5 × 4 = 20 (very probable × hospital treatment).

The same electrical flex trailing on the floor close to the rear wall of a rarely visited plant room might be rated as 1 × 4 = 4 (extremely unlikely × hospital treatment).

Note that in both instances the severity of injury is the same. This will sometimes be the case when the same hazard is being considered, but not always. For example, put the trailing flex in an old persons' care home and the foreseeable injury becomes more severe simply because the elderly have brittle bones and suffer severe injuries when they fall over.

There is no one right or wrong way to **calculate the risk**. Different organisations use different numbers and descriptions of likelihood and severity - some use a risk matrix, where numbers are used to describe the likelihood and severities, others use a 'low, medium, high' categorisation. It is the general principle that is important here, not the exact words and meanings.

The following graphic demonstrates how risk levels can be categorised using numbers and colour coding. In this example, green identifies a low risk, red identifies a high risk and intermediate risks are shown in between.

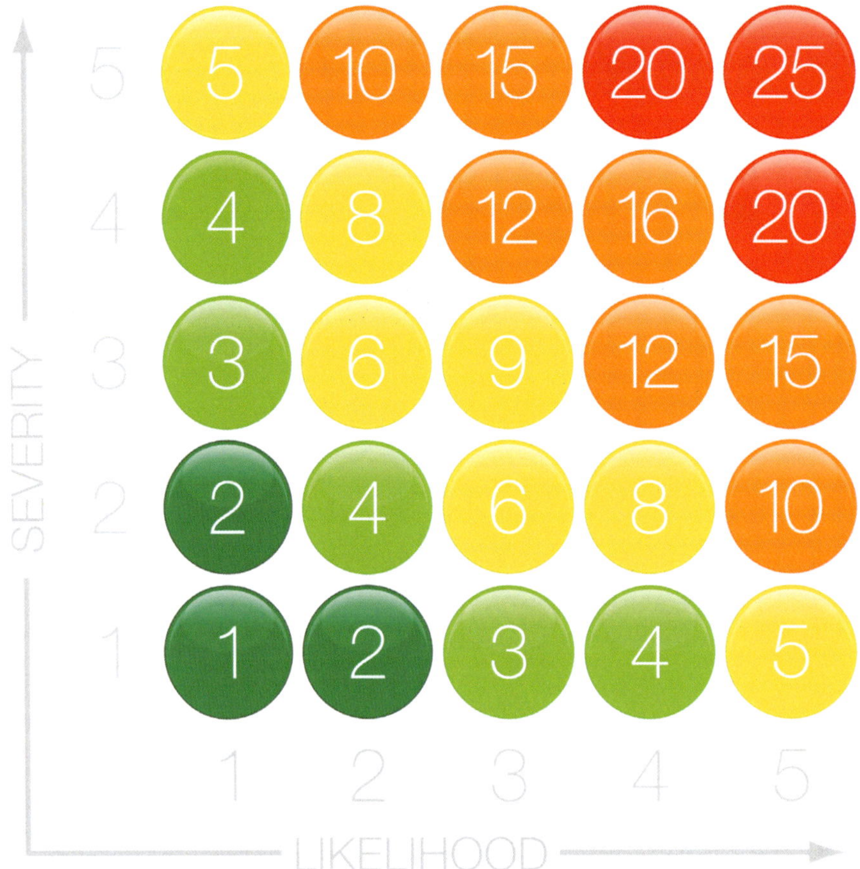

Risk assessment matrix

Using a risk-rating system such as our example matrix above can be useful for several reasons:

- **Clarity of thinking** - people tend to think more carefully about likelihood and severity of foreseeable injury when they are asked to use this type of scoring system, giving a more accurate end result.
- **Consistency of approach** - different people can use this system and will get similar results.
- **Prioritisation** - since risk is now represented by a number, and the higher the number the greater the risk, it is possible to easily separate out the various risks presented by several hazards and rank them in order.

Principles to Consider when Controlling Risk

Regulation 4 of **MHSWR** states that employers must implement control measures on the basis of the principles of prevention detailed in Schedule 1 of **MHSWR**. **CDM Regulations 2015** also require duty holders to apply the same principles. In descending order of preference, those principles are:

- **Avoiding risks**. If risks are avoided, there is no residual risk to manage. The best opportunity to avoid risk is at the design stage hence the duty imposed on designers.
- **Evaluating the risks which cannot be avoided**. Carry out a risk assessment to determine the full extent of the problem.
- **Combating the risks at source**. Replacing an old, noisy piece of equipment with a modern, quieter plant is more effective than issuing workers with hearing protection.

- **Adapting the work to the individual**. Especially the design of workplaces, the choice of work equipment and the choice of working and production methods. The main aim of these is to alleviate monotonous work and work at a predetermined work-rate, and reduce their effect on health.
- **Adapting to technical progress**. In recent times the construction industry has introduced technology that has significantly reduced risks for workers. For example the use of remote cameras - 'moles' - to carry internal inspection of vessels and sewers.
- **Replacing the dangerous by the non-dangerous or the less dangerous**. In many construction applications, volatile organic compound (VOC) paints have been replaced by acrylic paints which are water-soluble but become water-resistant when dry.
- **Developing a coherent overall prevention policy**. To effectively control risks, the whole safety system needs to be considered. This includes the individual, task, plant and equipment, the organisation, the management of the whole project, and the wider environment. It is important not to simply focus on the immediate hazards that are common to the construction industry; there is also a need to identify the underlying factors that cause injuries.
- **Giving collective protective measures priority over individual protective measures**. Securing a guard rail on a scaffold to prevent falling protects every one and requires no input from employees to be effective. Providing a harness only protects the user and is heavily reliant on the worker to ensure its effectiveness.
- **Giving appropriate instructions to employees**. Giving appropriate information in a clear and comprehensible form. This can be achieved through regular team meetings and toolbox talks.

Practical Application of the Principles - Applying the General Hierarchy of Control (ISO 45001:2018)

When hazards are identified through the risk assessment process, it is necessary to decide on the precautions needed to control those hazards to an acceptable level. This is the most important part of the risk assessment - identifying the further action that is needed and taking that action. When trying to decide what further precautions might be appropriate to a particular situation, a useful approach can be to use a **risk control hierarchy**.

TOPIC FOCUS

The general hierarchy of control (based on ISO 45001):

- Elimination.
- Substitution.
- Engineering controls.
- Administrative controls.
- PPE.

Each element is outlined below.

Elimination

If a hazard can be **eliminated**, then the risk created by that hazard disappears. This might be done by completely **avoiding** an activity that gives rise to risk. For example, an assembly workshop could stop welding steel in order to avoid the risks inherent in welding operations, and could buy in prefabricated metal components. The obvious limitation to this approach is that it is not possible to apply it to most of the activities carried out in the workplace. In this case, it may be possible to eliminate one or more hazards inherent in an activity. For example, hazardous substances can sometimes be replaced with materials that do the same job but present no risk to health (i.e. they are non-hazardous). Lifting equipment such as hoists and lifts can be used to completely eliminate manual handling. Machinery can be purchased that generates less noise to such an extent that there is no risk of hearing damage.

Substitution

Sometimes, hazard elimination cannot be achieved, but it is possible to **substitute** one hazard with another that creates less risk. For example, one hazardous substance classified as 'toxic' (i.e. lethal in small doses) is substituted with one that is 'irritant'. The replacement substance is still hazardous, but far less hazardous. A handling aid such as a sack truck does not eliminate manual handling, but it does reduce the risk of injury associated with moving boxes around in a workplace.

Engineering Controls

Engineering controls involve the use of an engineering solution to prevent exposure to the hazard.

This might be done by:

- **Isolation or total enclosure** - the aim here is to isolate the hazard physically so that nobody is exposed to it. This might be done by total enclosure or containment of the hazard, e.g. total enclosure of a process which generates dust to prevent its escape, acoustic enclosure of a noisy machine to reduce the noise exposure of those nearby, or guards around moving machinery to prevent contact.
- **Separation or segregation** - simply placing the hazard in an inaccessible location. An example would be overhead wires where an electrical conductor has been placed out of reach. In this case, precautions have to be taken to ensure that safe distances are maintained at all times (e.g. the use of goalposts to warn plant operators on a construction site of the safety distances for live electrical overheads).
- **Partial enclosure** - for example, a hazardous substance might be handled in a fume hood or partial enclosure which the worker can reach into for handling purposes. Air is extracted from the top or back of this partial enclosure so that any airborne contaminant is extracted from the enclosure away from the worker.
- **Safety devices** and features that ensure that the item is used in the correct way and not an unsafe way. For example, interlock switches are fitted to movable guards on machinery to ensure that, when the guard is open, the machine will not operate (but when the guard is closed, it will).

Administrative Controls

Administrative controls are those that rely on procedures and behaviour, such as:

- **SSOWs** - this is a formal procedure which defines a method of working that eliminates hazards or minimises the risks associated with them. SSOWs are necessary whenever hazards cannot be physically eliminated and some element of risk remains. This applies to any task involving significant risk. There is therefore a specified routine for setting and detonating explosives in a quarry. The safe system is essential to prevent accidents or other incidents. Certain high-risk work activities may be controlled by a permit-to-work system as a part of the SSOW.
- **Reduce exposure** - if the degree to which a worker is exposed to a hazard can be reduced, then that worker is far less likely to have an accident with that hazard. For example, an engineer who spends all day working on machinery with hazardous moving parts is more likely to suffer injury than an engineer who only spends one hour of their working day exposed to the same hazard. The duration of each exposure, (e.g. for 10 minutes or for 8 hours) and the frequency of exposure, (e.g. once a week or 10 times a day) will both play a part here - the less time and the less frequently, the better.
- **Reduce time of exposure** - many health hazards in the workplace cause a degree of harm that is entirely dependent on the **dose** that a worker receives, e.g. the harm caused by noise, vibration, radiation and most hazardous chemicals (such as lead). The dose is determined by two principal factors:
  - The concentration, intensity or magnitude of the hazard present.
  - The time of exposure.

For example, the harm to hearing caused by exposure to loud noise is entirely determined by the noise intensity (measured in decibels) and the duration of exposure:

- If you are exposed to the same noise intensity for twice as long, it gives you twice the dose of noise; if you are exposed for half as long, it gives you half the dose.
- The dose of noise determines the degree of damage done: the greater the dose, the more harm done.

  In all of the cases where harm is dose-related, limiting the time of exposure is an important control measure that can be practically used in the workplace.

- **Information, instruction, training and supervision**

  Training is instrumental in enabling employees to become competent. A competent employee is equipped with all relevant information and is fully aware of the hazards and the use of appropriate preventive measures.

  One way that an employer might provide basic health and safety information is through the use of safety signs (see below).

TOPIC FOCUS

Safety Signs

Safety signs combine shape, colour and pictograms to convey specific health and safety information or instructions.

The **Health and Safety (Safety Signs and Signals) Regulations 1996** divide safety signs into five categories:

- Prohibition - directed at stopping dangerous behaviour, e.g. 'No smoking'. The signs are circular with a black pictogram on a white background with a red border and red diagonal cross bar.
- Warning - tell people to be careful of a particular hazard, e.g. 'Forklift trucks operating in the area'. The signs are triangular with a black pictogram on a yellow background with a black border.
- Mandatory action - instruct people to take a specific action, often relating to wearing PPE, e.g. 'Eye protection must be worn'. They are circular with a solid blue background and a white pictogram.
- Safe condition - identify safe behaviour or places of safety, e.g. 'First-aid station'. They are rectangular or square with a white pictogram on a green background.
- Fire-fighting equipment - identify particular items of equipment, e.g. 'Hose reels'. They are rectangular or square with a white symbol or pictogram on a red background.

Prohibition
'No unauthorised entry'

Warning
'Toxic material'

Mandatory
'Safety boots must be worn'

Safe Condition
'Emergency escape route'

Fire equipment
'Hose Reel'

Examples of safety signs

Pictograms must be used on the signs, not just text. This is to overcome any language barrier that might be created if text alone were used (as a result of illiteracy, learning difficulties, language or certain eyesight impairments).

It is important to recognise that workers may not know the meaning of the safety signs in use in a workplace (some are obvious but others not) and therefore safety signs must be included in induction or job-specific training.

Supervision refers to management routinely checking workers and exercising their authority to control behaviour. Supervision, which is of critical importance as a management control, does not necessarily mean constant oversight of workers and the workplace. It is possible to supervise workers by making occasional contact with them at suitable intervals throughout a working period and it is possible to supervise workers remotely (i.e. from a distance).

Personal Protective Equipment

DEFINITION

Personal Protective Equipment

Equipment or clothing that is worn or held by a worker that protects them from one or more risks to their safety or health.

Worker wearing various types of PPE

There are instances where none of the above control measures can be used and there are times when some of them can but residual risk still remains. If this is the case, then it may be necessary to use PPE. Many different types of PPE are available, such as:

- Ear defenders for noise.
- Gloves to prevent contact with substances hazardous to the skin.
- Respiratory protection against substances hazardous by inhalation (breathing in).
- Eye protection against splashes of chemicals and molten metals, mists, sprays and dusts, projectiles and radiation including laser lights.

TOPIC FOCUS

Under the **Personal Protective Equipment at Work Regulations 1992**, it is the duty of the employer to:

- Supply suitable PPE where risks cannot be controlled by other more effective methods. 'Suitable' means:
  - It is appropriate for the risks and the conditions.
  - It is ergonomic (i.e. user-friendly).
  - It fits properly.
  - It does not increase overall risk.
  - It complies with any relevant standards.
- Ensure that when two or more items of PPE have to be worn together they are compatible.
- Provide suitable storage accommodation for PPE.
- Provide information, instruction and training to workers on the PPE they are expected to wear.
- Enforce the use of PPE.
- Replace or repair damaged or lost items.

The merits and limitations of PPE are summarised in the table below:

Merits	Limitations
• Can be used as an interim control whilst more expensive or difficult controls are put in place. • In some situations, it may be the only control option available. • It may be needed as a backup for emergencies when other controls have failed. • It is usually cheap. • It gives immediate protection.	• It only protects one person - the wearer. • It may not protect adequately if it is not fitted correctly. • It may not be comfortable and may interfere with the wearer's ability to do the job. • It may increase overall risk by impairing the senses (e.g. goggles that mist up). • It may not be compatible with other items that have to be worn or used. • People do not like wearing PPE. • If it fails, it fails to danger (the worker is exposed to risk).

Why a Hierarchy?

The controls options explained above are set out as a **hierarchy**: eliminating the hazard is the most preferred option and PPE the least preferred. The reason for this relates directly back to the human factors we discussed earlier in this element.

Workers do not behave in an ideal way in the workplace - they break rules knowingly and are subject to human error. Administrative controls and PPE are very reliant on personal behaviour and therefore are likely to be the least effective of all of the control options.

Application Based on Prioritisation of Risk

In Element 1 you will have covered the principle of 'so far as is reasonably practicable'. It is this principle that now forms the basis of our decision-making. When deciding on appropriate control measures, risks and risk ratings must be prioritised. Some of the factors to consider are:

- legal requirements;
- cost of implementation;
- level of risk;
- severity of consequences; and
- societal attitudes.

Use of Guidance; Sources and Examples of Legislation

The risk rating system (risk = likelihood × severity) described earlier is very useful as a practical tool for the day-to-day management of risk in a fast-changing workplace. It allows for a simple and consistent approach to the management of risk and the decision-making process. The one thing that this system does not allow for, however, is consideration of legal standards. If there are clear legal standards about the controls that should be applied to a particular hazard in a workplace, then the use of a scoring system and risk prioritisation and timescales, etc. becomes largely redundant. The only question that really matters is: are we doing what the law requires?

In some instances, there is a very clear legal standard that can be applied to the work in question. For example, the **Lifting Operations and Lifting Equipment Regulations 1998 (LOLER)** require that a thorough examination is carried out on lifting equipment. A risk assessment on a passenger lift should identify that the lift must be subject to a thorough examination once every six months, or in accordance with a scheme of thorough examination developed by a competent engineer in order to comply with **LOLER**.

However, in other instances, there is no clear legal standard, or the legal standard is open to interpretation, perhaps because the phrase 'reasonably practicable'" is included. In this case, there will probably be ACoPs and/or guidance published by the HSE and other authorities that clearly identifies the controls that are expected. Reference to the law, any semi-legal ACoP and guidance is, therefore, important in determining what the precautions should be for a particular hazard in the workplace:

- **ACoPs** have special legal status; follow the ACoP and you will be doing enough to comply with the law. Fail to follow the ACoP and you have to prove that you did something else that was at least as good.
- **Guidance** has no legal status but can be very useful in understanding what the legal standard is and how to achieve it.

Applying Controls to Specified Hazards

All hazards are different and bring with them their own characteristics. Even similar hazards will have different levels of risk depending on the circumstances. For example, a trailing cable across the floor of a room which is used occasionally for training might be adequately dealt with by using cable covers. However, the same trailing cable across the floor of a busy office or in an area where there are members of the public will require far more robust measures.

Residual Risk; Acceptable/Tolerable Risk Levels

Once control measures have been introduced and are taken into account, the current risk level can be estimated using 'Likelihood' and 'Severity.'

The risk that remains once controls have been taken into account can be referred to as the **residual risk**:

- If the residual risk is low, then it might be considered **acceptable** - the controls are adequate. Nothing more needs to be done.
- If the residual risk is high, a decision has to be made about whether this residual risk is tolerable or unacceptable:
  - **Tolerable** implies that it is not acceptable but it can be tolerated for a short time with interim controls put into place.
  - **Unacceptable** implies that the risk level is too high for work to be allowed in any circumstances.

In the case of both tolerable and unacceptable risk, additional controls will need to be put in place to reduce the risk down to an acceptable level. Once these controls have been implemented, a new residual risk level is created.

The HSE guide *Reducing risk and protecting people*, often referred to as R2P2, deals with this subject in great detail.

Distinction between Priorities and Timescales

A straightforward link can be made between the level of risk associated with a particular hazard and the prioritisation of that hazard: the higher the risk, the higher the priority.

You might also assume that a high risk demands a short timescale for corrective action and a low risk can be allowed to persist for a longer period of time, but the problem with this simple link is that an inspector might not accept that a long timescale could be applied to a simple, low-cost remedy; it should be implemented immediately because it can be done immediately.

So, priority and timescale are linked but they are not the same thing. Priority is the relative importance or urgency of an issue and will usually be linked to the risk level. Timescale is the length of time given for corrective action and must be decided based on the risk level and cost, practical difficulty, etc. of the control measure.

Step 4 - Recording Significant Findings

The significant findings of a risk assessment should be recorded to provide a statement of the hazards in the workplace, the extent of the risks that they present and the action taken to control those risks.

There is no standard format for risk assessments, so different organisations can adopt a format that is most appropriate to their circumstances. Typical content would include:

- Identification of the activity/area assessed and of the significant hazards.
- Identification of groups at risk and those especially at risk.
- Evaluation of the risks and the adequacy of existing control measures.
- Action plans for implementing further precautions needed.
- Date of assessment and name of the competent person carrying out the assessment.
- Review date.

Step 5 - Reasons for Review

A risk assessment conducted under **MHSWR** must be reviewed and amended as necessary if there is reason to suspect that it is no longer valid or if there has been a significant change.

TOPIC FOCUS

Triggers for a risk assessment review:

- **Significant change** to a matter that the risk assessment relates to:
  - Process.
  - Substances.
  - Equipment.
  - Workplace environment.
  - Personnel.
- There is **reason to suspect that the assessment is not valid**:
  - Accident.
  - Near miss.
  - Ill health.
  - Change to legal standards.

It is also good practice to review risk assessments on a regular basis. This is often done by determining a frequency of review based on the level of risk associated with the activity in question. An annual review of risk assessments is common practice in many workplaces.

Application of Risk Assessment for Specific Types of Risk and Special Cases

Examples of When They Are Required

The general approach to risk assessment described earlier in this chapter is suitable for the majority of hazards that are encountered in conventional workplaces but some hazards have their own characteristics. This is reflected in the fact that some areas of work have their own specific legislation, each of which require different risk assessment methods. Most of these subjects are dealt with elsewhere in the course but here are some examples:

- **Regulatory (Reform Fire Safety) Order 2005**
- **Display Screen Equipment Regulations 1992**
- **Manual Handling Regulations 1992**
- **Control of Substances Hazardous to Health Regulations 2002**
- **Control of Noise at Work Regulations 2005**
- **Control of Vibration at Work Regulations 2005**

Why Specific Risk Assessment Methods Are Used for Certain Risks

Specific risk assessments allow employers to systematically analyse the unusual - sometimes unique - hazards that arise. Subsequently they allow for the identification of specific control measures.

Special Case Applications

There are times when a risk assessment has to focus on one person or one specific group of workers because they are more vulnerable to particular hazards (or more at risk).

Young People

A young person is defined in **MHSWR** as anyone under the age of 18 years. The **MHSWR** require that they be given special consideration through the risk assessment process.

There are several reasons why a young person might be more vulnerable to risk in a workplace:

- Lack of experience in workplaces.
- Physical and mental immaturity.
- Poor perception of risk.
- Heavily influenced by peer group pressure.
- Eager to show a willingness to work.

Young people are especially vulnerable in workplaces

For these reasons, you often need to think more carefully about the work that a young person is doing. It may be necessary to:

- Prohibit a young person from carrying out certain high-risk activities (e.g. operating high-risk machinery).
- Restrict their work patterns and hours (no nightshift work or overtime).
- Train and supervise them to a greater degree than other workers.
- Provide mentors to monitor and supervise young persons more closely than other workers and to provide clear lines of communication.

Expectant and Nursing Mothers

Pregnant women and women who have just given birth (whilst breastfeeding or for the first six months after pregnancy) are more at risk from certain types of hazard. In most of these instances, the hazard presents a risk not only to the woman but also to the baby. Many of these hazards can cause miscarriage, birth defects or ill health in the baby. Consequently, new and expectant mothers are also picked out by **MHSWR**, which require that they be given special consideration through the risk assessment process.

TOPIC FOCUS

Hazards that present greater risk to pregnant women:

- Certain hazardous chemicals (e.g. lead).
- Certain biological agents (e.g. the rubella virus).
- Manual handling, especially later in pregnancy.
- Extremes of temperature.
- Whole-body vibration.
- Ionising radiation.
- Night shift work.
- Stress.
- Violence.

In all cases, where a woman reports that she is pregnant, a risk assessment should be carried out focusing on the work that she is doing and the hazards that might increase risk to her and the child. It may then be necessary to:

- Change the type of work or the way that it is done.
- Change the hours of work.
- Suspend the woman from the workplace on full pay.

There are other employment and anti-discrimination laws that should be considered, but that is beyond the scope of this course.

New and expectant mothers are given special consideration by MHSWR

MORE...

Further information on new and expectant mothers can be found at:

www.hse.gov.uk/mothers/index.htm

Disabled Workers

People with disabilities may be at greater risk from particular hazards, depending on the nature and extent of their disability. For example, a visually-impaired worker may be able to carry out a packaging operation at their workstation without any risk to themselves or others, but they may find it very difficult to evacuate from the building during a fire using an escape route that is unfamiliar to them. In this instance, they may need some assistance in the form of an 'evacuation buddy'.

During the risk assessment process, it may be necessary to:

- Identify certain health and fitness criteria for some jobs and then screen staff against these criteria. This may have the effect of excluding those with a certain disability from doing these jobs, (e.g. forklift truck drivers should have their eyesight checked before being allowed to drive).
- Identify workers with known disabilities and consider what the implications of their particular type and level of disability might be.

Again, remember that there is other employment and anti-discrimination legislation that should be considered at the same time as any relevant health and safety law.

MORE...

Further information on disability and health and safety can be found at:

www.hse.gov.uk/disability/index.htm

Lone Workers

People who work entirely on their own for periods of time, or those who are not alone but are not with colleagues whom they can rely on for help, might be classified as lone workers. For example, a service engineer who spends four hours alone in a plant room servicing machinery is a lone worker; but so is a healthcare worker who travels around in the local community visiting elderly patients to provide care (even though they may be in sight of other people at all times during their working day).

Lone workers are a group of workers who are especially vulnerable in certain instances:

- They may be more at risk of violence, particularly if the worker is exposed to members of the public, has to travel out into the community or is involved in work that brings them into contact with violent people, (e.g. prison staff or mental health nurses).
- They may be more at risk if they are injured or fall ill. Certain types of work involve a high risk of personal injury or ill health, (e.g. confined space entry). In these situations, lone working may be inappropriate or additional precautions may be necessary to protect the individual.

STUDY QUESTIONS

14. Why is the distinction between hazards and risks so important to health and safety management?
15. Identify the purpose and objectives of risk assessment.
16. What techniques are used for identifying hazards?
17. What are the five categories of health hazard?
18. Identify the five steps involved in risk assessment.
19. Apart from office workers, what particular staff groups require special consideration during a risk assessment in an office environment?
20. What factors are used to evaluate risk?
21. What is residual risk?
22. State the general hierarchy of control.
23. What conditions might trigger a risk assessment review?
24. When should PPE be used?

(Suggested Answers are at the end.)

Summary

This element has introduced you to improving health and safety culture and assessing risk.

In particular, this element has:

- Defined safety culture as the shared attitudes, values, beliefs and behaviours relating to health and safety, explained how this culture may be positive or negative and described the influence of peers
- Discussed the three human factors that influence a worker's behaviour: organisational, job and individual factors.
- Outlined the influence of a worker's attitude and motivation on their safety-related behaviour and considered the importance of risk perception.
- Looked in detail at some of the issues that must be dealt with in order to improve safety culture:
  - Clear management commitment with visible leadership.
  - Competent staff (training, knowledge, experience and other skills).
  - Communication of safety information in verbal, written or graphic form and through health and safety media.
  - Consultation.
  - Training at appropriate times, e.g. induction training for new staff.
- Defined the terms 'hazard', 'risk' and 'risk assessment'.
- Identified the aim of risk assessment - to ensure that hazards are eliminated or risk is minimised by the correct application of relevant standards.
- Explained the approach to risk assessment:
  - Identify the hazards.
  - Identify the people who might be harmed and how.
  - Evaluate the risk and decide on precautions.
  - Record the significant findings and implement them.
  - Review and update as necessary.
- Outlined the groups of people who might be harmed, such as employees, contractors, visitors and members of the public.
- Discussed the principle of rating risk using a simple scoring system for likelihood and severity.
- Considered the importance of legal standards in the risk evaluation process and the idea that a general risk control hierarchy can be used to reduce residual risk to an acceptable level.
- Discussed the need to focus on vulnerable groups of workers such as young persons, expectant and nursing mothers, disabled workers and lone workers so as to further their protection.

Exam Skills

Question

Scenario

You have been appointed as the competent person to assist the site manager in undertaking their duties under health and safety legislation. You decide to carry out a site inspection focussing on site behaviour and practices. You notice general housekeeping is poor, personnel not wearing PPE or following site rules. In addition, you observe the site foreman not enforcing site rules and allowing bad practices to continue. On challenging the foreman about this, he smiles, shrugs his shoulders and says "this is a building site, this is how things are done around here".

Task: Improving Health and Safety Culture

You decide to prepare a brief on how you intend to improve the safety culture on the site. Prepare some notes on the key aspects you would cover in order to change the culture and bad practices you have seen.

(Total: 10 marks)

Approaching the Question

Now think about the steps you would take to answer this question:

Step 1 The first step is to **read the scenario carefully**. Note in the scenario you have observed a number of examples of bad practices and identified from discussing safety with the foreman that safety is not considered particularly relevant to the work being carried out.

You decide to provide a 'briefing document' so will need to structure your approach using the four headings given.

Step 2 Now look at the **task** - prepare some notes on key areas that would require to be undertaken to change the safety culture.

Step 3 Next, consider the **marks** available. In this task, there are 10 marks available, so it is expected that around four key headings of information should be provided. Each piece of information will require explanation carrying a further 2-3 marks per heading, giving a total of 10 marks.

Step 4 **Read the scenario and task again** to make sure you understand it and have a clear understanding of how you can improve safety culture. (Re-read your study text if you need to.)

Step 5 The next stage is to **develop a plan** - there are various ways to do this. Remind yourself, first of all, that you need to be thinking about "improving safety culture" and what techniques can be used to achieve this. The answer plan will take the form of a bullet-pointed list that you need to develop into a full answer.

Suggested Answer Outline

Headings for the answer would fall into the following areas:

(a) Management commitment and leadership – visible leadership.

(b) Competent workers.

(c) Effective communication – verbal, written and graphic.

(d) Co-operation and consultation – safety committees, safety forums and safety circles.

Now have a go at the question yourself.

Example of How the Question Could be Answered

Improving the safety culture on site and, as a result, reducing the frequency of accidents will require taking an approach following the principles of the 4 Cs, Control – Competence – Communication – Co-operation and Consultation.

'Control' refers to management commitment and leadership where visible leadership can be demonstrated by behaving safely, e.g. following site rules, involvement in the day-to-day management of health and safety by attending safety meetings and taking part in safety tours and audits.

Competence is about ensuring all personnel are competent in the roles they are expected to perform. This means ensuring everyone understands their job role and has received appropriate training and, where necessary, ongoing training to maintain competence. This includes training in health and safety.

Effective communication from management to workers and vice versa should be established through the organisation ensuring correct information is transmitted, received and understood. The means of delivery of information is through verbal, written and graphic mediums.

Co-operation and consultation can only be created in an organisation with effective worker co-operation and involvement. If workers feel they are being dictated to then they will feel little ownership of health and safety. Involving workers in the decision-making process is the best way of achieving this. Examples of how this can be done could be through safety committees, safety forums, safety circles and toolbox talks.

Reasons for Poor Marks Achieved by Exam Candidates

- Not following a structured approach for the briefing document; failing to identify the four key ways in which safety culture can be improved.
- Not expanding on the four key headings of the 4 Cs. The numbers of marks awarded would probably be reduced to 2-3 marks.
- Not providing an answer to the task question at all. If you do not attempt all tasks, you cannot get any marks for the task questions you omit.

Element 3

Managing Change and Procedures

Learning Objectives

Once you've studied this element, you should be able to:

1. Outline the procedures that contribute to how change can be managed effectively and reviewed.
2. Define a safe system of work and explain how safe systems are developed, used and monitored for general work activities.
3. Define a permit-to-work system and explain why, how and when permits to work are used.
4. Explain why emergency procedures are needed, how these procedures are planned, their impact on fire evacuation and first aid, and the need for their continual review.
5. Outline the four steps to follow when investigating incidents and understand how to record and report fatalities, injuries, diseases and dangerous occurrences.

Contents

Managing Change

IN THIS SECTION...

- Managing change in the workplace impacts health and safety.
- There are a number of factors that influence the impact of change such as communication, risk assessment, competency, segregation of work areas, amendments of emergency procedures and welfare provision.
- Change needs to be constantly reviewed to ensure that the construction plan is being kept on track.

Typical Changes Faced in the Workplace and Their Possible Impact

In construction projects, changes are very common and likely to occur at any stage of the project.

Most changes, if not managed properly through a formalised process, will have considerable impact. Managing change effectively is crucial to the success of a construction project.

Common examples of change that can impact on health and safety are:

- Construction works.
- Change of process.
- Change of equipment.
- Change in working practices.

Managing the Impact of Change

Communication and Co-operation

People do not change simply because they receive an email or attend a training program. The message must be communicated clearly to all of those who are affected. This can be done through regular toolbox talks and team meetings.

Risk Assessment

As the nature of the work changes, for example from demolition through site clearance to the construction phase, then the hazards and risks will change accordingly. Regular review of risk assessments is essential to ensure that they remain relevant.

Appointment of Competent People

The **Management of Health and Safety at Work Regulations 1999 (MHSWR)** require employers to appoint a competent person. Similarly, the **CDM Regulations 2015** require clients to be satisfied with the competence of the other duty holders that they appoint. One of the main reasons for these requirements is that professional advice is readily available to assist with the identification of new hazards that are introduced at various stages of the project and that they are dealt with correctly.

Segregation of Work Areas

Construction sites change rapidly, new processes and new plant and equipment are introduced regularly. Traffic routes and work areas are redefined accordingly. Likewise, segregation of these areas has to be reviewed and updated to accommodate these changes.

Amendment of Emergency Procedures

Emergency arrangements must be regularly kept under review and amended when necessary. For example, as building work progresses and the site layout changes, fire exit routes and muster points might need to be repositioned.

Welfare Provision

As changes in the construction process occur, so too will the numbers and types of workers. The number and location of welfare facilities must be reviewed. For example, there could be a change in process which requires showers and decontamination rooms to be provided.

Review of Change (During and After)

The review process plays an important part in the success of any management system. Regular review meetings - which might involve the client, contractors, site management and even in some cases, local residents and other interested parties - help to ensure that the construction plan is kept on track.

One of the main reasons for a review meeting on completion of the project is to learn lessons not only to ensure that the same mistakes are not repeated but also to identify the successes so that similar processes can be used in future.

STUDY QUESTIONS

1. How can the impact of change be managed?
2. Why is a review system important for measuring the success of change to a workplace system?

(Suggested Answers are at the end.)

Safe Systems of Work for General Work Activities

IN THIS SECTION...

- A Safe System of Work (SSOW) is a formal procedure based on a systematic examination of work in order to identify the hazards. It defines safe methods of working which eliminate those hazards or minimise the risks associated with them.
- It is the responsibility of the employer to develop SSOWs with the involvement of both competent persons and employees who will be carrying out the work. These safe systems must be documented.
- SSOWs are normally developed using the process of task analysis, which involves breaking work down into a series of steps so that hazards can be identified and risks controlled at each step using technical, procedural and behavioural controls. Once developed, safe systems must be implemented and monitored to ensure continued effectiveness.

Introduction to Safe Systems of Work

An SSOW is a formal procedure based on a systematic examination of work in order to identify the hazards. It defines safe methods of working which eliminate those hazards or minimise the risks associated with them.

We can identify three key elements from this definition of an SSOW:

- The SSOW is **formal** - documented or recorded in some way.
- It results from **a systematic examination of work in order to identify the hazards** - it is the result of risk assessment.
- It **defines safe methods** - it is the safe procedure or work instruction.

So, simply put, the employer should carry out a systematic risk assessment, identify the hazards and precautions necessary, and then formally record the safe way to carry out the task taking this all into account.

TOPIC FOCUS

To be effective, the SSOW must bring together these elements or factors in such a way as to create a safe work method known by the abbreviation **PEME** :

- **People**: Who is the SSOW for? What level of competence or technical ability should they have? Are there vulnerable persons involved? How many people are working on the task?
- **Equipment**: What plant or equipment will be worked on? What equipment will be used? What safety equipment will be required? What are the hazards associated with the job and the equipment?
- **Materials**: What materials will be used or handled during the work? What are the hazards of the materials? How will waste be disposed of?
- **Environment**: In what type of environment will the work take place, e.g. space, light and temperature? Might any of these factors increase risk or need to be controlled?

Why Workers Should be Involved When Developing Safe Systems of Work

The competent person must work closely with the employees who will be doing the work. The workers involved should take an active part in all stages of both the development and review of SSOWs. Their practical knowledge and skills provide a valuable source of information about the nature of the risks, including unusual ones, and methods of working. They can also contribute by assessing plans and written documentation, and providing feedback on the effectiveness of the system in practice.

Involvement in this way enables employees to gain a deeper understanding of the hazards and risks, and of the way in which the SSOW will minimise those risks. This helps to give ownership of the safe working methods to workers and to develop a positive safety culture.

Why Procedures Should be Recorded/Written Down

Documenting SSOWs provides a precise reference for all workers, and ensures consistency of method. It also provides a reference for use in training and instruction in safe procedures. The recording of SSOWs may be in the form of short notes, or perhaps manuals detailing exactly what steps to take when carrying out more complex and lengthy procedures, such as calibrating and setting up grinding wheels. SSOW documentation can be accompanied by checklists for employees to use as aids to ensure that all the correct steps are taken, and to tick off details before continuing with the next step or starting operations. Written SSOWs also provide the employer with a written record which may be required for legal reasons.

The Differences between Technical, Procedural and Behavioural Controls

An SSOW will involve all the elements of control that we identified earlier in the general hierarchy of controls:

- **Technical, or engineering controls** - applied directly to the hazard in order to minimise the risk. This may involve fencing or barriers of different kinds to isolate workers from the hazard as far as possible, or fail-safe devices designed into equipment to stop its operation if there is a fault. It includes controlling environmental issues such as lighting and the use of Local Exhaust Ventilation (LEV) to control dust.
- **Procedural controls** - the way in which work should be carried out in relation to the hazard. They will specify the exact tasks involved, their sequence, and the safety actions and checks which have to be taken. Often, procedures will relate to the correct operation of technical controls, for example, procedures for the use of safety equipment such as Personal Protective Equipment (PPE) and machine guards, and for the reporting of problems.
- **Behavioural controls** - address how the individual worker acts in relation to the hazard. They include training, supervision and peer observations.

Developing a Safe System of Work

SSOWs are developed by task analysis, prior to work commencing, as a part of the planning process. Task analysis is the process of breaking a job down into its component steps and then identifying the hazards associated with each step. A safe working method can then be identified to deal with each hazard.

Analysing Tasks, Identifying Hazards and Assessing Risks

Can you remember what the acronym **SREDIM** represents? We outlined it earlier in Element 2 when we discussed hazard identification methods and task analysis. It means:

- **S**elect the task to be analysed.
- **R**ecord the steps or stages of the task.
- **E**valuate the risks associated with each step.
- **D**evelop a safe working method.
- **I**mplement the safe working method.
- **M**onitor to ensure it is effective.

So, for example, a vehicle breakdown and recovery company might perform task analysis on the job of changing the wheel on a customer's car.

This analysis might identify the key steps of the task as:

- **Step 1**: Park breakdown vehicle.
- **Step 2**: Remove faulty wheel.
- **Step 3**: Fit spare wheel.
- **Step 4**: Leave.

The hazards associated with each of these steps would have to be identified and the risk evaluated. For example, at Step 2, one of the hazards identified might be movement of the customer's car after it has been jacked up, causing it to collapse off the jack. Another hazard might be that of being struck by passing traffic. The risk of this happening would be greater if you were changing the wheel on the hard shoulder of a busy motorway.

Introducing Controls and Formulating Procedures

Once the risks have been evaluated, the appropriate safe working method can be developed. In our example, one of the controls would be for the worker to check that the customer's handbrake has been firmly applied and that the handbrake actually works. Another would be for the worker to wear high-visibility clothing at all times.

Implementing the SSOW is often more problematic than developing the safe working method, because implementation requires workers to adopt the new working procedures and use all the identified controls. As we have discussed in Element 2, people's behaviour can be difficult to control. One way of overcoming or minimising this difficulty is to consult with and involve workers in the development process so that they are able to raise objections and concerns at an early stage; they then have some ownership of the new methods.

Instruction and Training in How to Use the System

A key step in the implementation of any safe working method is the provision of information, instruction and training. In some cases, if the workers have the competence to interpret and correctly follow new methods, simply informing them of changes to existing methods will be sufficient. In other instances, detailed theoretical and practical training will have to be provided to ensure that workers understand and can apply the safe working methods.

Monitoring the System

The last step of the task analysis process is monitoring; once the safe working method has been put into place, it should be checked periodically. This is to ensure that the:

- New safe working method is being correctly followed and applied. If it is not, then supervision must be improved.
- Method is, in fact, safe. If it is not, then it will have to be reviewed and amended accordingly.

TOPIC FOCUS

What to consider when developing an SSOW:

- Details of the task or activity to be performed.
- Details of the equipment and materials to be used.
- Any information or guidelines provided by manufacturers.
- Number of employees who will carry out the activity and their level of competence and training.
- The history of any accidents associated with the activity.
- The adequacy of the control measures in place.
- Any relevant legal requirements.
- The need for consultation with employees regarding the SSOW.
- Any emergency procedures that may be required.
- Systems for, and level of, monitoring and supervision.

STUDY QUESTIONS

3. Define a safe system of work.
4. How does involving employees in the development of safe systems of work contribute to strengthening the safety culture?
5. What is the difference between technical, procedural and behavioural controls?
6. Why do instruction and training form a part of safe systems?

(Suggested Answers are at the end.)

Permit-to-Work Systems

IN THIS SECTION...

- Permit-to-work systems form part of an SSOW to control high-risk work activities, such as hot work.
- A permit system formalises the control of high-risk work to ensure that all the risks have been identified, all the precautions put in place and that appropriate information has been communicated to all relevant parties.
- A permit to work usually has four main sections:
  - Issue.
  - Receipt.
  - Clearance.
  - Cancellation.
- Permit-to-work systems rely on the use of paper permits, but ultimately they only control risk properly when correctly used.
- Permits to work are typically used to control hot work, work on non-live (isolated) electrical systems, entry into confined spaces, work at height and some forms of machinery maintenance work.

Meaning of a Permit-to-Work System

A permit-to-work system is a formal, documented safety procedure, forming part of an SSOW, which ensures that all necessary actions are taken before, during and after particularly high-risk work.

TOPIC FOCUS

The sort of high-risk work that would normally be controlled by a permit system includes:

- Hot work (involving naked flames or creation of ignition sources).
- Work on high-voltage electrical systems.
- Confined space entry.
- Work on operational pipelines.
- Excavating near buried services.
- Maintenance work on large, complex machinery.

The high-risk nature of the work is the key feature that these types of work have in common, which makes them subject to permit control. If the work is not carried out in precisely the right way then workers and others may be killed. It is the intention of the permit system to focus everybody's attention on the high-risk nature of the work to ensure that:

- The correct safety precautions are in place before, during and after the work.
- All the people who need to know about the work do actually know about it.

Why Permit-to-Work Systems are Used

A permit-to-work system is a management system that is used to help prevent accident or injury to personnel, prevent damage to plant/machinery and to prevent damage to a product or to a client site. They are used in situations where the risk is higher than normal. Examples are given later in this section.

For work of a very high risk nature, the Health and Safety Executive (HSE) has produced *Guidance on permit-to-work systems: A guide for the petroleum, chemical and allied industries* (HSG250).

How Permit-to-Work Systems Work and Are Used

Permits to work are formal documents specifying the work to be done, the hazards and the precautions to be taken. The permit provides a clear written record, signed by a responsible manager or supervisor, that all foreseeable hazards have been considered and all the necessary actions have been taken. It must be in the possession of the person in charge of the work before work can commence.

TOPIC FOCUS

The general details to be included in a permit to work reflect the operation of the permit system and can be summarised as:

Issue

- Description of the work to be carried out (details of plant and location).
- The date and time of issue, and the duration over which the permit will be open (remain valid).
- Assessment of hazards associated with the job.
- Controls required, such as isolations, PPE and emergency procedures.
- Signature of the authorised person issuing the permit.

Receipt

- Signature of the competent person accepting the permit.

Clearance

- Signature of the competent person stating that the area has been made safe (e.g. work completed) and that they are leaving the area and isolations can be removed.

Cancellation

- Signature of the authorised person stating that the isolations have been removed, the area has been accepted back and that the equipment can be re-started.

Each permit will have a unique identification number so that copies of a permit can be traced back to their specific source.

Permits are usually triplicate copy documents with a unique identification number for cross-reference purposes.

In essence, the permit identifies the hazards and controls needed to do the work safely and then passes the responsibility for the area from the site to the workers. On completion, the workers hand the area back to enable it to be checked and normal work to resume.

The sections of a permit to work operate in the following way.

Permit to Work in Confined Spaces or Dangerous Atmospheres

Site/Building/Area:.. Exact Location:...

Job Details:...

This permit is only valid when all sections are complete. If you are in doubt or don't understand, then please ask. Please insure that you sign this permit to work. So not proceed with your work until your permit has been authorised by the relevant member of staff.

By accepting this permit you agree the requirements of the Company Confined Spaces Code.

HAZARDS TO BE AWARE OF AND PRECAUTIONS TO BE TAKEN

	YES	NO	N/A
Are you qualified/trained to undertake this work?			
The confined space has been isolated from all copnnected pipework			
The confined space has been purged with steam/water/air			
The confined space is electricity isolated and locked out			
The confined space is mechanically isolated and locked off			
The confined space is below 30°c on full cooling			
Is the entrance big enough to allow access and egress in an emergency?			
Supply of respirable air assured/ventilation required?			
Means of access to and escape from the confined space acceptable?			
Is breathing apparatus at hand and in good working order?			
Is the safety line/tripod/harness and any other back up equipment to hand?			
Are there adequate emergency arrangements in place?			
Trained standby person at point of entry?			
Are you qualified/trained top undertake this work?			
The confined space has been isolated from all connected pipework			

ATMOSPHERE TEST REQUIRED

Time of Test:

Oxygen	%	Pass/Fail	
Carbon Monoxide	%	Pass/Fail	
Carbon Dioxide	%	Pass/Fail	
Flammability	%	Pass/Fail	
Other (specify)	%	Pass/Fail	

Other safety precautions required:

Other safety equipment required:

Type of power tools and lighting remitted:

PREPARATION COMPLETE. ACCEPTANCE AND AUTHORISATION

I verify the above location has been examined, the precautions on the ch3ecklist have been taken, and that permission is authorised for this work. I also accept responsibility for the work to be carried out.

Person responsible for work:.. Signed:..

Authorised Permit Issuer:... Signed:...

Date and Time:... Time of Expiry:...

Example of a permit to work

Issue

This section of the permit defines the work, identifies the hazards and determines the necessary safety precautions.

An authorising manager must complete this section. This will require them to carry out a risk assessment of the work in order to identify all the relevant hazards and precautions. The manager must be competent to do this.

The authorising manager must specify:

- The exact nature of the work.
- Where the work can take place.
- The names of each of the workers authorised to carry out the work.
- The date and time that work can start.
- The period of time the permit is valid for.
- The control measures that must be in place before, during and after the work.
- Any restrictions.
- Any other permits that may be relevant.

The authorising manager signs the permit to formally confirm that all necessary precautions have been taken and that work can now start, provided the necessary precautions are adhered to. The manager's name and signature, and the date and time, should be clear.

Receipt

Here, the handover process to allow work to start is formalised.

The workers sign the permit to formally confirm that they understand all the hazards, risks and precautions and that they will comply with all necessary control measures.

Names should appear clearly in capitals with signatures, dates and times.

Clearance/Return to Service

The workers sign this section of the permit to confirm that they have left the workplace in a safe condition, that work is complete and that normal operations may resume.

Cancellation

The authorising manager signs this section to accept the hand-back of the workplace from the workers. This also has the effect of cancelling the permit so that no further work can take place under its authority.

Extension

This section is included in some permit systems in case there is any overrun of the work. It allows the authorising manager to grant an extension to the timescale of the permit. If the permit is to be in operation for more than one shift, there must be a formal system to facilitate the handover of the permit between shifts.

Permits are often issued in triplicate:

- One copy is displayed in the area where the work is taking place.
- One copy stays with the authorising manager.
- One copy is displayed in a central location (often on a permit board) where other permits are also displayed for clear communication.

Remember that a permit to work is just a piece of paper; it does not ensure safety. What ensures safety is the management system that it represents. In some cases, permits are treated simply as unnecessary paperwork - to be filled in because someone at head office says so. This can encourage casual practices, such as authorising managers issuing permits without actually checking that control measures have been put in place, which can lead to unfortunate consequences.

Cutting with an oxy-acetylene torch requires a permit to work

A good permit system is only as good as the managers using it. To work effectively:

- Only authorised managers should issue permits.
- Precautions must be checked before permits are authorised.
- Permits must never be amended.
- The permit must be treated with respect.
- All permit conditions must be adhered to.
- Staff must be trained and competent.
- The system must be monitored to ensure it is effective.
- The permit-to-work system must be appropriate for the nature of the business, e.g. a bakery may require a less complex system than an oil rig.
- Sufficient time must be allowed to ensure permits are issued correctly.

When to Use a Permit-to-Work System

Hot Work

Permit systems are commonly used to control hot work where naked flames will be used (e.g. propane, butane or oxy-acetylene torches) or where a significant ignition source will be created (e.g. welding or grinding operations).

Typical precautions for control of hot work:

- Flammable materials are removed from the work area.
- Items that cannot be removed are covered with fire-retardant blankets.
- The floor is swept clean.
- The wooden floor is damped down.
- A suitable fire extinguisher is at hand.
- A 'fire-watcher' is present in the area.
- The work area is visited routinely after the work has finished to check the area for smouldering.

Work on Non-Live (Isolated) Electrical Systems

The high risk associated with working on, or near, electrical systems means that this type of work is usually subject to permit control. In particular, permits are usually required for work on, or near, high-voltage systems.

A permit system is used to ensure that:

- All precautions are in place.
- The workers are competent to do the work.

Working live is justified only under certain, specified circumstances one of which is if it is not possible to work with the power off. A different permit which deals with the far greater problems of live working must be introduced.

Machinery Maintenance

Maintenance work often involves the removal or disabling of safeguards and control systems. For large, complex industrial machinery, more than one person may be involved in the work and they may be required to work inside the machinery. This can generate high risk that might be best controlled using a permit system.

Machinery maintenance

A permit system is used to ensure that:

- Work is carefully planned, assessed and controlled.
- The nature of the work is communicated to those who need to know about it.
- Power sources are isolated and locked off.
- Stored energy is released or secured.
- The workers are competent to do the work.

Confined Spaces

Entry into confined spaces can be extremely hazardous, so should always be under the control of a permit-to-work system. This will require a competent person to carry out a risk assessment and then develop an SSOW that identifies all the necessary precautions for entry and the emergency arrangements that must be put in place.

Safe working in confined spaces (L101) is the Approved Code of Practice (ACoP) and guidance which accompanies the **Confined Spaces Regulations 1997** about how a permit system can be developed.

Work at Height

Some work at height may be controlled under a permit to work. This could detail the area where work is permitted, the method of access (ladder, stairs, etc.), safe access routes and fall arrest precautions to be taken.

STUDY QUESTIONS

7. What is a permit to work?
8. What are the four key elements of a typical permit?

(Suggested Answers are at the end.)

Emergency Procedures

IN THIS SECTION...

- An organisation should develop emergency procedures to deal with foreseeable incidents such as fire, bomb threat and chemical spill.
- These procedures should cover the internal arrangements for dealing with the foreseeable incidents, which will include:
  - The foreseeable emergencies.
  - Procedures for raising the alarm.
  - Procedures to follow.
  - Provision of suitable equipment.
  - Nomination of responsible staff.
  - Dealing with the media.
  - Contacting the emergency services.
- Employers must provide and notify staff of its first-aid facilities, equipment and personnel.
- Fire procedures must be in place so that employees are aware of what actions to take in the event of fire such as means of escape, emergency evacuation procedures, the role of fire marshals/wardens, fire drills and provisions for people with disabilities.
- Suitable emergency arrangements must be provided when working near water.
- Emergency procedures should be continually reviewed and updated during a build to reflect any changes and should consider the use of Temporary Accommodation Units (TAUs).

Why Emergency Procedures Need to be Developed

Despite all the controls that can be introduced into an organisation, things can still go wrong. Accidents and incidents do happen. When they do, it is critical that the organisation has emergency procedures that can be brought into effect without delay, otherwise there may be a poor or inappropriate response that makes things worse rather than better.

Clean up after oil spill

TOPIC FOCUS

An organisation has to develop procedures to deal with foreseeable incidents. Such incidents might include:

- Fire.
- Bomb threat.
- Spillage of a hazardous chemical.
- Release of a toxic gas.
- Outbreak of disease.
- Severe weather or flooding.
- Multiple casualty accident.
- Terrorist/security incident.

The foreseeable incidents will vary depending on many factors, such as the type of organisation and its location.

What Arrangements Must be Made When Planning Emergency Procedures and First-Aid Provisions

Regulation 30 of the **CDM Regulations 2015** states that suitable and sufficient arrangements for dealing with any foreseeable emergency must be made and, where necessary, implemented, and those arrangements must include procedures for any necessary evacuation of the site or any part of it.

In particular, account should be taken of the:

- type of work for which the construction site is being used;
- characteristics and size of the construction site, and the number and location of places of work on that site;
- work equipment being used;
- number of persons likely to be present on the site at any one time; and
- physical and chemical properties of any substances or materials on, or likely to be on, the site.

Those who could be affected must be informed of the procedure and the arrangements must be tested at suitable intervals. (This is discussed in more detail later in this section.)

In a general context, having identified the foreseeable incidents, the organisation should make internal arrangements to deal with each of them should they occur. These arrangements should include:

- **Foreseeable emergencies**: an outline of the foreseeable emergencies that might occur along with the circumstances in which they might occur and their foreseeable outcomes. This will identify what the emergencies are and why procedures need to be followed.
- **Procedures for raising the alarm**: workers need to know how the relevant alarms are raised for different types of emergency and any escalation procedure that might apply. For example, manual call points might be used to raise the alarm in the event of fire, but radio or telephone notification to a control centre might be used in the event of a security incident.

- **Procedures to be followed**: in the event of a fire, for example, normal practice is for workers to exit the building using the signed escape routes and assemble at a designated place. In the event of a bomb threat, the procedure is often the exact opposite: to go to a room inside the building away from windows and external walls.
- **Provision of suitable equipment**: if there is a chemical spill, for example, absorbent granules or booms might be used to contain the spill and PPE used to prevent harm to those involved in the containment operation. In the event of a release of toxic gas, respiratory protective equipment may be needed.
- **Nomination of responsible staff**: in a fire situation, there is likely to be a need for fire wardens or marshals who walk through the building to check that everyone is aware of the fire evacuation; a fire team may also be required whose job will be to check the area where the fire is suspected to be.
- **Dealing with the media**: it is foreseeable that some emergencies will attract media attention. Procedures for handling the media should be developed and for nominated staff trained in media handling techniques (such as how to answer questions from newspaper and radio reporters).
- **Arrangements for contacting emergency services**: while internal emergency arrangements must be made by the organisation to deal with foreseeable incidents, these responses will normally also involve contacting external emergency services for help. There should therefore be appropriate arrangements in place:
  - Communication equipment, e.g. landline and mobile phones, satellite phones or VHF radio. The more remote the location, the more difficult this becomes.
  - Contact details, e.g. national and local emergency service numbers. This may involve international medical evacuation as well.
  - Responsible individuals with the necessary information and knowledge nominated to make the call. In many instances, the emergency services can provide a more effective response if they are given critical information quickly.

First-Aid Provisions

An employer has a duty, under the **Health and Safety (First Aid) Regulations 1981**, to make appropriate first-aid provision for their employees. This is to allow an immediate emergency medical response to foreseeable injuries that might occur in the workplace. This provision consists of three elements:

- **Facilities** – an appropriate location where first-aid treatment can be given.
- **Equipment** – suitably stocked first-aid kits and other equipment as necessary.
- **Personnel** – staff with appropriate training to deliver first-aid treatment.

The employer must notify staff of these first-aid arrangements and, in particular, the identity of trained first-aid personnel.

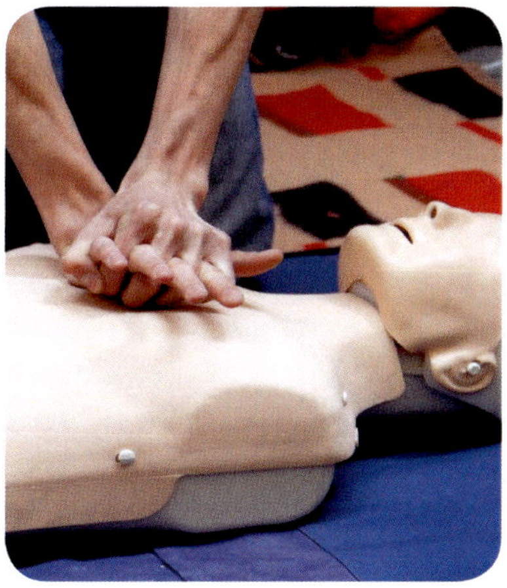

First-aid training

First-Aid Facilities

Suitable facilities should be provided where first-aid treatment can be given. As a minimum, this might consist of a room that is used for other purposes but can be quickly converted into a treatment area. In a larger workplace, a dedicated treatment room should be provided.

This room should be:

- Centrally located in an area that can be accessed by the emergency services.
- Clean and adequately heated, ventilated and lit.
- Provided with hand-wash facilities, a chair and a clinical waste bin, etc.

First-Aid Equipment

As a minimum, one fully stocked first-aid kit might be provided for a small, low-risk workplace.

TOPIC FOCUS

The HSE provides suggestions on what should be included in a first-aid kit, for example:

- Large triangular bandages (preferably sterile).
- Sterile plasters (individually wrapped; appropriate to the type of work).
- Disposable gloves.
- Safety pins.
- Sterile eye pads.
- Large- and medium-sized, unmedicated wound dressings (sterile; individually wrapped).

MORE...

Further information on first aid at work is available from:

www.hse.gov.uk/firstaid

In larger workplaces, multiple first-aid kits should be positioned at various locations in the workplace, plus:

- Eye-wash stations.
- Emergency showers.
- Blankets.
- Splints.
- Resuscitation equipment, such as Automatic External Defibrillators (AEDs).
- Stretchers.
- Wheelchairs.
- Other equipment, as required.

Eye-wash station

The contents of a first-aid kit are covered by the ACoP and guidance to the regulations. Small first-aid kits can be provided for drivers and lone workers.

First-Aid Personnel

Role of First Aiders

The basic role of first aiders is to keep the injured casualty alive until professional medical assistance can take over. This is sometimes referred to as the 'three Ps':

- Preserve life.
- Prevent deterioration.
- Promote recovery.

First aid is also concerned with the provision of simple treatment to minor injuries that do not require professional treatment (such as issuing a plaster to a worker with a very minor laceration).

Types of First-Aid Personnel

Minimum provision would be a person available to take charge of the first-aid arrangements, which means looking after the first-aid equipment and facilities, and calling the emergency services when required. Such **appointed persons** should be available at all times while people are at work. Appointed persons are not necessary where there are adequate numbers of first aiders. Appointed persons do not need to be trained as first aiders.

A step up from the appointed person is workers trained to **Emergency First Aid at Work (EFAW)** level. The number of appointed persons and/or emergency first aiders will depend on a first-aid needs assessment.

Where the needs assessment shows there is a higher level of risk, first aiders trained to the **First Aid at Work (FAW)** standard should be appointed in numbers appropriate to the risk and numbers of workers.

EFAW and FAW personnel must have suitable training and hold a valid certificate of competence. EFAW courses are usually one-day long. FAW courses usually run over three days. Re-qualification training for these courses usually takes one and two days, respectively.

The HSE also advises that EFAW and FAW personnel should have annual refresher training to prevent 'skills-fade', although this is not compulsory.

HSE guidance also states that the employer may decide to make provision for mental ill health by training some workers to allow them to identify and provide support to someone suffering mental ill health (sometimes referred to as 'mental health first aid').

Selecting Staff to be First Aiders

There are a number of factors that should be taken into account when selecting persons to fulfil the role of a first aider at work. These include:

- Being reliable, having a good disposition and good communication skills.
- Awareness of their own limitations and limitations of the training.
- Having the aptitude and ability to absorb new knowledge and learn new skills.
- Having the ability to cope with stressful situations and accept responsibility.
- Physically fit enough to deal with the demanding nature of some aspects of first aid.
- Consideration of the need for first aiders considering gender, ethnicity and religious convictions.
- The person's normal duties should be such that they can leave their work immediately at any time to attend an emergency.

First-Aid Coverage

The Principal Contractor should ensure that they undertake a first-aid needs assessment in order to determine what first-aid facilities, equipment and trained personnel to provide.

There should be adequate first-aid cover on all shifts worked in the workplace, with weekend, morning, afternoon and night shift work patterns all adequately provided for.

TOPIC FOCUS

Factors to consider when deciding first-aid provision and number of first aiders:

- The general risk level of the workplace.
- The hazards present in the workplace.
- Accident history.
- The presence of vulnerable persons.
- The number of workers in the workplace.
- Work patterns and shift systems of workers.
- Provision for non-employees (e.g. shoppers, schoolchildren).
- The geographic location of the workplace.
- Distribution of the workforce (e.g. number of buildings on a site, number of floors, etc.).
- The physical size and spread of the workplace.
- The need to cover holidays and absence.
- Peripatetic employees (those who travel from place to place).

The geographic location is an important issue to consider when determining first-aid provision. Workplaces within easy reach of the emergency services can perhaps provide minimal cover, but workplaces in remote locations, that the emergency services may take hours to reach, must be far more self-reliant.

MORE...

Further information on first aid at work is available from:

www.hse.gov.uk/firstaid

Leaflet INDG347(rev2), *Basic advice on first aid at work*, HSE, 2011, available at:

www.hse.gov.uk/pubns/indg347.pdf

Principles of Fire Evacuation

Fire procedures ('action in the event of fire') must be in place to instruct employees in what actions to take in the event of fire. They need to know what to do if they find a fire, or hear the alarm (indicating someone else has found a fire). The safety message is simple - get out and stay out!

Means of Escape

DEFINITION

MEANS OF ESCAPE

The safe route that a person takes from wherever they happen to be in a building or on site, to a place of safety outdoors.

The main requirement of an escape route is that everyone is able to escape to a place of safety. In order to evacuate a fire situation safely, there must be one or more escape routes available for people to use and the following principles can be applied:

Escape routes should be kept clear of obstructions

- A means of escape is to be available for every person in the workplace, whether they are in an office, workshop, out on a construction site or on scaffolding.
- The means of escape should allow an able-bodied person to travel the entire distance of the route alone and unaided.
- They should not have to use machinery (e.g. a passenger lift) except in special circumstances (when it must be rated for escape purposes).
- The means of escape must take the person from wherever they are in the workplace to a place of safety outdoors.
- Two or more escape routes may have to be available in case one becomes blocked or otherwise unusable.
- The distance that a person has to travel along the route from its start to the final exit out of the building (or the assembly point on site) should be as short and straight as possible and may be limited to a maximum distance.
- The width of corridors and passageways in buildings and walkways on site should allow the free and fast movement of the numbers of people who might be anticipated (there are minimum widths applied).
- Escape routes should be clearly signed and appropriately lit.
- Escape (emergency) lighting should be provided where necessary in case the mains power fails.
- All escape routes should be clear of obstructions (stored materials, vehicles or inappropriate doors).

The means of escape and emergency plans must be regularly reviewed as construction work progresses to ensure they are always suitable and sufficient. As structures change, routes may become blocked or may no longer be the best route for occupants to take.

Stairs and Passageways

Stairs and passageways (corridors) used as means of escape should be:

- Protected against fire ingress to a higher degree than other parts of a building. (This is to keep them free from flames and smoke so they can be used for escape. The walls, floor and ceiling will need to be fire-resistant and doors along the route should be fire doors.)
- Kept free from obstructions and the storage of materials and products that would be fuel in a fire.

Doors

Doors along the escape route must:

- Be suitably fire-resistant for the location (see earlier).
- Be easily opened by a person escaping.
- Be wide enough to allow unimpeded passage.
- Open in the direction of travel (not always a strict requirement where occupancy numbers are low).
- Be able to be opened at all times while people are in the building (i.e. not locked or chained while the building is occupied).

Lighting

Escape routes must be adequately lit, and where normal lighting is not adequate or will be lost in a power failure, additional lighting may be necessary. In the simplest of situations this could be a rechargeable torch, but larger workplaces will require escape lighting. This should:

- Illuminate:
  - The escape route.
  - Fire safety signs and equipment.
- Be maintained in working order.
- Undergo routine testing.

Signs

Escape routes must be easy to follow. Fire safety signs should be provided to direct occupants to and along routes. These signs (called 'safe condition' signs) must:

- Comply with legislative standards.
- Have appropriate pictograms in addition to any wording.
- Be fixed in places where they are easily seen and not easily obstructed.

Signs can be standard coloured signs, light-retentive (glow in the dark) or attached to escape light units.

A fire escape sign on an escape light unit

Assembly Points

Escape routes lead to an **assembly point**, where people gather once outside a building or away from danger areas on site. It will be in the open air and away from any further danger from the fire. At this location a roll call is taken to ensure everyone has escaped safely.

Assembly points should be:

- Clearly signed.
- A safe distance from a building or site hazards.
- In a safe location (away from traffic routes and storage areas).
- At a location where further escape is possible if needed.
- Out of the way of emergency service vehicles attending the fire and their personnel.

The location should be continually reviewed as construction work progresses.

In some cases a '**refuge**' is provided inside a building, which is a temporary location in a place of comparative or relative safety, in a fire-protected area, where people may wait for a short time, e.g. people with impaired mobility waiting for assistance to evacuate further.

TOPIC FOCUS

Travel Distances

- The distance people need to go to escape (the travel distance) should be as short as possible. The travel distance should be measured from the farthest point in a room to the door to a protected stairway or, if there is no protected stairway, to the final exit from the building.
- If there is only one escape route, the travel distance should not normally be more than 18 metres. This distance should be shorter (12 metres or less) in any parts of the premises where there is a high chance of a fire starting or spreading quickly.
- The distance can be longer (up to about 25 metres) where the chance of a fire starting or spreading quickly is very low.
- If there is more than one escape route, the travel distance should not normally be more than 45 metres (around 25 metres in areas where the risk of fire is high and about 60 metres in areas where the risk of fire is very low).

Emergency Evacuation Procedures

- Every construction workplace must have procedures to ensure the safe evacuation of people from danger areas in the event of fire.
- Some procedures will require nominated persons to carry out certain duties, e.g. fire marshals or someone to take the roll call.
- Employees should be trained in fire procedures and nominated persons trained in their specific roles.
- Fire drills allow employees to practise their emergency response and help management to monitor the effectiveness of the emergency arrangements.
- Special procedures may be required to ensure safe evacuation of the infirm or disabled.

Role and Appointment of Fire Marshals/Wardens

However many employees there are, responsibility for action in the event of fire should be assigned to specific persons. Some premises may have fire marshals (or fire wardens) who are responsible for the following actions:

- Ensuring all occupants leave by the designated safe escape routes.
- Searching all areas to ensure that the area is clear (people have left).
- Ensuring that fire escape routes are kept open and clear at all times.
- Ensuring all doors and windows are closed on leaving the area.
- Conducting the roll call at the assembly area (in the absence of senior management).
- Meeting the fire service on arrival and informing them of any relevant details.

- Ensuring special assistance is available to the disabled or infirm (perhaps using 'evac-chairs' in multi-storey areas).
- Investigating the location of the fire (as indicated by the fire alarm system controls).

Some sites may operate a 'fire team' who have the dual role of investigating the fire alarms and fire-fighting (they will be specially trained for this).

Fire Drills

Emergency procedures should be practised through drills and exercises to ensure that people are familiar with the actions they might be expected to take. In this way, people's responses become automatic. For example, fire evacuation drills should be conducted routinely in all workplaces and multiple casualty accident exercises should be practised in high-risk workplaces where they are a foreseeable event.

Workers will only know what to do when these various emergencies occur if they have been provided with information and training. Any nominated individuals will require additional training on their roles in the emergency and on the safe handling of any equipment (e.g. PPE) that they might have to use. Members of the public may require information on emergency procedures, which might be provided in the form of notices, or by means of public address system announcements.

Provisions for People with Disabilities

An evacuation plan for disabled people is commonly known as a PEEP: a Personal Emergency Egress Plan or Personal Emergency Evacuation Plan. The PEEP explains the method of evacuation to be used by a disabled person in each area of a building.

Disabled people who require their evacuation to be facilitated will need to be advised of the options available to them. In some instances, they will need to be given assistance to evacuate. Those disabled people who require assisted evacuation should have a pre-arranged meeting place. If the disabled person is likely to move around the building, a communication process will be necessary between the appointed 'buddy' and the disabled person so they can arrange to meet at a particular location.

For further information, the HSE has published *Fire safety risk assessment: Means of escape for disabled people*.

Suitable Emergency Arrangements When Working Near Water

Regulation 26 of the **CDM Regulations 2015** states:

"Where, in the course of construction work, a person is at risk of falling into water or other liquid with a risk of drowning, suitable and sufficient steps must be taken to:

- *prevent, so far as is reasonably practicable, the person falling;*
- *minimise the risk of drowning in the event of a fall; and*
- *ensure that suitable rescue equipment is provided, maintained and, when necessary, used so that a person may be promptly rescued in the event of a fall."*

Typical situations include:

- working on bridges above waterways;
- working in marinas;
- working in cofferdams and caissons; and
- working on suspended scaffolds above water.

Suitable arrangements would include the provision of personal buoyancy equipment, trained rescuers, stop nets or lines, and rescue nets/boats/line. Arrangements should also be made to practise emergency situations. If a worker falls into the water, then the primary concern is to keep them afloat, make sure their location is immediately obvious and tracked, and effect a rescue as soon as possible.

Continual Review of Emergency Procedures as a Build Continues

The emergency procedure should be reviewed whenever there are changes to the physical layout of a site, or changes to processes, materials and personnel, and the required updates made.

There should also be a review after each exercise or drill, or after an actual emergency when the plan was put in motion to assess if there are areas that need to be improved.

Inclusion of TAU within the Emergency Plan

All workplaces must afford safe access and egress. This is a particular concern should an emergency arise. TAUs may change location as construction continues - they may be inside other existing buildings. TAUs often house welfare facilities and sleeping accommodation which may hinder a worker's ability to evacuate quickly.

STUDY QUESTIONS

9. What is the main objective of an emergency procedure?
10. What types of incidents may require the development of an emergency response?
11. What factors might need to be considered when determining the first-aid provision for a workplace?
12. What is the purpose of signs used along an escape route?
13. Outline the main requirements for an escape route.
14. What is an assembly point, and how might it differ from a refuge?

(Suggested Answers are at the end.)

Learning from Incidents

IN THIS SECTION...

- There are different levels of incident investigation ranging from minimal to high depending on the circumstances.
- Four steps are employed in the procedure of incident investigation:
  - Step 1: Gather factual information about the event.
  - Step 2: Analyse that information and draw conclusions about the immediate and root causes.
  - Step 3: Identify suitable control measures.
  - Step 4: Plan the remedial actions.
- 'Reportable' incidents include fatalities, specified injuries, diseases and dangerous occurrences.
- Records of work-related injuries must be reported and recorded, for example, in a standard accident book.
- Certain types of incidents have to be reported to the enforcing authorities in a specific way to a timescale to comply with statutory requirements of the **Reporting of Injuries, Diseases and Dangerous Occurrences Regulations 2013 (RIDDOR)**.

The Different Levels of Investigations (HSG245)

The HSE's *Investigating accidents and incidents* (HSG245) provides a 'ready-reckoner' type matrix to assist in determining the level of investigation that is required. When deciding, you must consider the worst possible outcome. A serious near miss such as a scaffold collapse may not have caused injury but had the potential to cause multiple fatalities.

Minimal

The supervisor looks at the circumstances to try to learn lessons that will prevent a similar incident in future.

Low

Involves a short investigation by the supervisor or line manager to identify the immediate, underlying and root causes.

Medium

Involves a more detailed investigation by the supervisor or line manager and may involve the safety adviser and employee representatives.

High

This is a team-based approach. It will involve the same people as a 'medium' investigation but it will be carried out under the supervision of senior management or directors.

Basic Incident Investigation Steps

When investigating an accident or other types of incidents, there are some basic principles and procedures that can be used:

- Step 1: Gather factual information about the event.
- Step 2: Analyse that information and draw conclusions about the immediate and root causes.
- Step 3: Identify suitable control measures.
- Step 4: Plan the remedial actions.

However, before the investigation can begin, there are two important issues that should be considered:

- **Safety of the scene** - is the area safe to approach? Is immediate action needed to eliminate danger even before casualties are approached?
- **Casualty care** - any injured people will require first-aid treatment and may need hospitalisation. This is, of course, a priority. It is also worth considering the welfare of uninjured bystanders who may be in shock.

Once immediate danger has been eliminated and casualties have been attended to, a decision may have to be made about the type or level of investigation. Is this to be a:

- Relatively simple investigation of an incident that caused no or minor outcomes and did not have the potential to cause serious outcomes?
- More in-depth and thorough investigation of an incident with serious outcomes or potential for serious outcomes?

The first type of investigation might be carried out by the line manager of the area; the second type often involves a team of investigators that might include a safety specialist, senior managers, a technical specialist and perhaps worker representatives.

Checklists can be useful tools to aid an investigation to ensure that the relevant points have been covered.

TOPIC FOCUS

Items that could be included on an accident investigation checklist could include:

- Personal details of the person involved.
- Time and location of the accident.
- Type and severity of the injury sustained.
- Whether the injured person had been given first aid, had returned to work or had been sent to hospital.
- The underlying medical condition of the injured person.
- The task being undertaken at the time of the accident.
- The working environment as far as weather, standard of lighting and visibility were concerned.
- The condition of the floor or ground.
- The type and condition of any PPE that was being worn.
- Details of the training and information received.
- Details of any relevant risk assessments that had been carried out.
- Any previous similar accidents that had occurred.

Step 1 - Gathering the Information

Factual information about the event can be gathered by:

Accident investigator taking measurements

- Securing the scene as soon as possible to prevent it being altered.
- Collecting witnesses' details quickly, before they start to move away. In some cases, it may help to remove witnesses from the scene and ask them to wait in a separate area. If there are many witnesses, it may be better to separate them from each other to prevent collusion or contamination of their testimony.
- Collecting factual information from the scene and recording it. This might be done by means of:
  - Photographs.
  - Sketches.
  - Measurements.
  - Videos.
  - Written descriptions of factors, such as wind speed, temperature, etc.
  - Taking physical evidence such as samples, or the equipment that has failed.
  - Marking up existing site/location plans.

  The investigator should come prepared with the appropriate equipment to record this information.
- Once the scene has been thoroughly examined, moving on to the second source of information: witnesses.

  Witnesses often provide crucial evidence about what occurred before, during and after incidents. They should be interviewed carefully to make sure that good-quality evidence is gathered.

TOPIC FOCUS

Good witness interview technique requires that the interviewer should:

- Hold the interview in a quiet room or area free from distractions and interruptions.
- Introduce themselves and try to establish rapport with the witness using appropriate verbal and body language.
- Explain the purpose of the interview (perhaps emphasising that the interview is not about blaming people).
- Use open questions, such as those beginning with 'What?', 'Why?''?, 'Where?', 'When?', 'Who?', 'How?', etc. that do not put words into the witnesses' mouths and do not allow them to answer with a 'yes' or 'no'.
- Keep an open mind.
- Take notes so that the facts being discussed are not forgotten.
- Ask the witness to write and sign a statement to create a record of their testimony.
- Thank the witness for their help.

Once witnesses have been interviewed, move on to the next source of information: documentation.

Various documents may be examined during an accident investigation, such as:

- Company policy.
- Risk assessments.
- Training records.
- SSOWs.
- Permits to work.
- Maintenance records.
- Site plans.
- Area layout plans.
- Previous accident reports.
- Sickness and absence records.

Step 2 - Analysing the Information

The purpose here is to draw conclusions about the **immediate** and **root** causes of the incident.

Immediate causes are the obvious causes that gave rise to the event itself. These will be the things that occurred at the time and place of the accident. For example, a worker slips on a patch of oil spilt on the floor, injuring his back as he falls backwards and hits the ground. The immediate cause of the back injury is hitting the ground, but there are many contributors to this cause. It is common to think of these in terms of unsafe acts and unsafe conditions. So here, for example, we might have the slippery oil (unsafe condition), and the worker walking through it (unsafe act).

Root causes are the things that lie behind the immediate causes. Often, root causes will be failures in the management system, such as:

- Failure to adequately supervise workers.
- Failure to provide appropriate PPE.
- Failure to provide adequate training.
- Lack of maintenance.
- Inadequate checking or inspections.
- Failure to carry out proper risk assessments.

For example, with the slip described above, the root causes might be a poorly maintained machine that has leaked oil onto the floor, and a poorly inspected and maintained workshop with broken light fittings and inadequate lighting levels. Here, the worker might be blameless on the basis that, given those conditions, the accident was bound to happen eventually.

Many of the accidents that happen in workplaces have one immediate cause and one underlying or root cause. The root cause gives rise to the immediate cause which in turn gives rise to the accident (rather like a row of dominoes falling; in fact, this idea is often referred to as the domino theory of accident causation).

If that one root cause is identified and dealt with, then the accident should not happen again. For example, if a worker twists their ankle in a pothole in the pavement then the obvious solution is to fill the pothole in. That deals with the immediate cause. It would also be worth asking how long the pothole had been there for. If it had been there for a long time, why was it not spotted sooner? And if it had been spotted, why had it been left unrepaired with no interim measure being taken to protect people?

These questions might identify an underlying cause, such as inadequate inspection and maintenance, or failure to put interim measures in place while waiting for maintenance work to be carried out. These root causes need to be addressed if similar accidents are to be prevented in future.

In contrast to this single cause idea, some workplace accidents are complex and have multiple causes: there are several immediate causes for the accident and each of these has one or more underlying or root cause. This idea is usually referred to as multi-causation theory. For example, a worker might be struck by a load being carried by a forklift truck.

Immediate causes for such an accident might be:

- Failure to secure the load on the pallet.
- Poor road positioning of the truck close to a pedestrian exit.
- Aggressive braking by the truck driver.
- An inattentive pedestrian stepping out in front of the truck.

On investigation, each of these immediate causes might have their own separate **root causes**, such as:

- No training for the driver, who is new to the workplace, has not worked with this type of load before and is unaware of the load-securing technique required.
- Lack of segregation of pedestrian and traffic routes; no barriers and no marking to separate the two.
- Lack of proper driver induction into their new workplace so they are unaware of the layout and position of pedestrian exits, etc.
- Poor maintenance of the truck.
- No refresher training for existing staff, meaning that experienced staff have become complacent.

If there are multiple causes for the accident, then it is important that each of these causes is identified during the investigation, otherwise incomplete remedial action will be taken and similar accidents may happen in the future.

Step 3 - Identify Risk Control Measures

Once the immediate and underlying causes of the accident are known, appropriate control measures can be identified. It is important that the correct control measures are established, otherwise time, money and effort will be wasted on inadequate and unnecessary measures that will not prevent similar occurrences in the future.

Control measures must be identified to remedy both the immediate and underlying causes. Immediate causes are usually easy to identify - if there is a spill of oil on the floor, clean it up; if the guard is missing from the machine, re-attach it.

Underlying causes can be harder to determine because they reflect failure of the management system. But it is essential that the correct control measures to remedy the failure of the management system are identified, because this will help prevent similar accidents occurring in similar circumstances across the entire organisation. For example, if a worker slips on some oil that has leaked out of a vehicle in the distribution depot, an employer may:

- Clean up the oil leaking out of the vehicle (the immediate cause), but fail to deal with the underlying cause (lack of inspection and maintenance). This could lead to more leaks, which in turn would lead to more pedestrian slips (and perhaps, vehicle skids).
- Clean up the oil leaking out of the vehicle and deal with the underlying cause (by introducing a proper inspection and maintenance system). In this instance, there is a good chance that most oil leaks will be prevented in the future for all vehicles in the fleet at all locations.

Perhaps the most important questions to ask when identifying control measures are:

- If this action is taken, will it prevent this same accident from happening in exactly the same way at this location?
- If this action is taken, will it prevent other similar types of accidents from happening in similar locations in the future?

If the answer to both of these questions is 'no', then you need to identify other control measures.

Step 4 - The Action Plan and its Implementation

An accident investigation should lead to corrective action being taken in just the same way as a workplace inspection will. Remedial actions can be presented in an action plan as illustrated below:

Recommended Action	Priority	Timescale	Responsible Person
Introduce induction training for all new drivers	Medium	1 month	Warehouse manager

When the action plan is being prepared, appropriate immediate and interim control measures must be given suitable priorities and timescales. Unsafe conditions must not be allowed to persist in the workplace. Dangerous practices and high-risk activities must be dealt with immediately. This means that immediate action must be taken to remedy these circumstances when they are discovered. Machinery and equipment may have to be taken out of action, certain work activities suspended and locations evacuated. These responses cannot be left until the investigation has been completed. They will have to be implemented immediately to ensure safety while the investigation is in progress.

There may be interim control measures that can be introduced in the short- to medium-term to allow work to proceed while longer-term solutions are being worked out. For example, hearing protection might be introduced as a short-term control measure until the maintenance of a piece of machinery that is producing excessive noise has been completed. A perimeter guard might be fitted around an overheating machine that would ordinarily be protected with a fixed enclosed guard while new cooling units are sourced and delivered.

Underlying causes will often demand significant time, money and effort to remedy. It is essential, therefore, that the remedial actions that will have the greatest impact are prioritised and timetabled first. There may be actions that have to be taken (to address a management weakness, or to achieve legal compliance) that will not be as effective in preventing future accidents. These actions should still be taken, but with a lower priority.

TOPIC FOCUS

The **contents of a typical incident investigation report** may include:

- Date and time of the incident.
- Location of the incident.
- Details of the injured person/persons involved (name, role, work history).
- Details of injury sustained.
- Description of the activity being carried out at the time.
- Drawings or photographs used to convey information on the scene.
- Details of witnesses and witness statements.
- Immediate and underlying/root causes of the incident.
- Assessment of any breaches of legislation.
- Recommended corrective action, with suggested costs, responsibilities and timescales.
- Estimation of the cost implications for the organisation.

How Fatalities, Specific Injuries, Diseases and Dangerous Occurrences Must be Recorded and Reported

DEFINITIONS

REPORTING

The process of informing people that an incident has occurred - this can be internally within the organisation or externally to enforcing authorities or insurers, etc.

RECORDING

The process of documenting the event.

Work-related incidents, including near misses, should be reported internally by workers to management. The system put in place by an organisation to allow for this should be described in the 'arrangements' section of the organisation's safety policy.

It is standard practice for workers to report incidents to their immediate line manager verbally, followed by completion of an internal incident report form that records the incident.

There are occasions when this simple verbal reporting procedure is not appropriate and a more complex reporting procedure then has to be introduced. For example, a lone-working contractor visiting a client's premises may have to report their accident to the client as well as to their remote line manager.

Recording and the Accident Book

When a work-related incident is reported, a record is usually created of that event (in some instances, the report is filed in written form, so reporting and recording are one and the same thing).

As a minimum, organisations must keep a record of all work-related accidents that result in personal injury.

This is dictated by social security legislation and there is a standard accident book that can be used. The accident book provides a record on any injury (or disease) which could give rise to a claim for Industrial Injuries Disablement Benefit. Without such a record, the benefits can be harder to obtain.

There is no fixed format for an accident report form. Organisations can use any format, as long as certain standard information is recorded

This accident book contains a form that is completed to make a record of each accident. This record must then be removed from the book and kept by a responsible person under lock and key in order to comply with **The Data Protection Act 2018** which implements the European Union's **General Data Protection Regulations** (which seeks to ensure that personal information is only kept and used for legitimate purposes and that access to that personal information is restricted to those with a legitimate need). These records have to be kept for a minimum of three years.

Organisations do not have to use the standard accident book. They can design their own in-house forms, which is often better to take account of near misses and other incidents that may not be covered on the standard form. But these forms must contain the basic information that is collected by the standard accident book as a minimum if they include accident reporting.

TOPIC FOCUS

Typical contents of an accident record include:

- Name and address of casualty.
- Date and time of accident.
- Location of accident.
- Details of injury.
- Details of treatment given.
- Description of event causing injury.
- Details of any equipment or substances involved.
- Witnesses' names and contact details.
- Details of the person completing the record.
- Signatures.

Separate forms can also take account of near misses and reports of ill health. These do not need to record standard accident book data as there is no explicit legal requirement to keep a record of most near misses.

Organisations often have separate forms for the recording of accidents (as above) and the recording of accident investigations. This is an important distinction to make; the accident record is the initial record of the basic facts of the injury; the accident **investigation** report is the detailed examination of what caused that injury and why it happened (immediate, underlying and root causes) as well as the corrective actions required to prevent recurrence. These do not have to be recorded in the same document.

Internal Incident Reporting

When establishing an incident-reporting policy, the organisation should be clear about the type of incidents that have to be reported by workers. It is usual to include a list of definitions in the policy so that workers understand the phrases used. If the organisation wants workers to report near misses, it must specify this in the policy and be clear about what the phrase 'near miss' actually means.

Having established an incident-reporting policy, the organisation must encourage workers to report all relevant incidents. Unfortunately, there are many reasons why workers do not report incidents.

TOPIC FOCUS

Barriers to reporting - reasons why workers might not report accidents:

- Unclear organisational policy on reporting incidents.
- No reporting system in place.
- Culture of not reporting incidents (perhaps due to peer pressure).
- Overly complicated reporting procedures.
- Excessive paperwork.
- Too time consuming.
- Blame culture (where a worker feels that they may be treated detrimentally or disciplined for reporting any accident or near miss that they were associated with).
- Belief that management does not take reports seriously.
- Concern over the impact on the company or departmental safety statistics (especially if this is linked to an incentive scheme).
- Reluctance to receive first-aid treatment.
- Apathy.

The organisation should try to remove each of these barriers to ensure that every relevant incident is reported in a timely manner. Most of these barriers can be dealt with by having a well-prepared, clearly stated policy, adopting user-friendly procedures and paperwork, and training staff in the procedures. An organisation can take disciplinary action against workers who fail to report incidents if they have been given the training and means to do so.

Sometimes, it is not sufficient simply to report an incident to immediate line management; sometimes, a higher-profile response is necessary. When fatal or major injuries, high-cost events, high-profile incidents or environmental events occur, it will be necessary to notify certain internal personnel immediately. Directors, senior managers, human resources managers, health and safety and/or environmental managers, and worker representatives may all have to be notified. Action by these staff may then be required to inform external parties as necessary, e.g. the family of the casualty, external authorities, insurance companies, public relations advisers, etc. These internal and external contact procedures, or **escalation procedures**, should be documented in the incident-reporting section of an organisation's safety policy.

The Reporting of Injuries, Diseases and Dangerous Occurrences Regulations 2013

Reportable Events

RIDDOR require the reporting of certain types of events (reportable events) to the relevant enforcing authorities.

A nominated responsible person should be identified by the employer for carrying out **RIDDOR** reporting requirements. This would usually be done in the 'arrangements' section of the health and safety policy.

There are six main types of reportable events.

Fatality

When a person dies as a result of injury from a work-related accident (with the exception of suicide but including an act of physical violence), if death occurs within one year of the original injury, it must be reported even though it may have been reported earlier under a different category.

Specified Injuries

A worker suffering from specified injuries listed in Regulation 4 of **RIDDOR** including:

- A fracture, other than to fingers, thumbs and toes.
- Amputation of an arm, hand, finger, thumb, leg, foot or toe.
- Permanent loss of sight or reduction of sight.
- Crush injuries leading to internal organ damage.
- Serious burns (covering more than 10% of the body, or causing damage to the eyes, respiratory system or other vital organs).
- Scalpings (separation of the skin from the head) which require hospital treatment.
- Unconsciousness caused by head injury or asphyxia.
- Any other injury arising from working in an enclosed space, which leads to hypothermia, heat-induced illness, or requires resuscitation or admittance to hospital for more than 24 hours.

Dangerous Occurrences

Dangerous occurrences are events that fall within the list in Schedule 2 of **RIDDOR**. They do not have to have caused a reportable injury and are categorised as near miss events. The 27 categories of dangerous occurrence include:

- The collapse, overturning or failure of load-bearing parts of lifts and lifting equipment.
- Plant or equipment coming into contact with overhead power lines.
- Explosions or fires causing work to be stopped for more than 24 hours.

Occupational Disease

This is where a worker suffers one of the eight categories of reportable occupational diseases (listed in Regulations 8 and 9):

- Carpal tunnel syndrome.
- Severe cramp of the hand or forearm.
- Occupational dermatitis.

- Hand-arm vibration syndrome.
- Occupational asthma.
- Tendonitis or tenosynovitis of the hand or forearm.
- Occupational cancer.
- Any disease attributed to an occupational exposure to a biological agent.

To be reportable, these diseases must be diagnosed, in writing, by a medical practitioner and they must arise from a specific work exposure as stated in the Regulations.

Over-Seven-Day Injuries

An employee or self-employed person who is away from work or unable to perform their normal work duties for more than seven consecutive days (not counting the day of the accident) falls into this category.

Immediate Hospitalisation of a Non-Worker

This category of reportable event is where a work-related injury involves a member of the public (e.g. a pensioner in a shop) or people who are not at work (e.g. a pupil at a school). They only need reporting where the person is injured in a work-related accident and is then taken from the scene of the accident to hospital **for treatment** to that injury. A key phrase here is 'for treatment' - diagnostic tests like X-rays don't constitute treatment.

Reporting Procedures

The following explains how events should be reported:

- Most incidents, including **fatalities, specified injuries and dangerous occurrences** must be reported by the 'quickest practicable means' without delay. This can be done using an online notification report form.
- **Fatalities and specified injuries** can be reported by telephone to the HSE Incident Contact Centre.
- For all of the above incidents, an online report must be submitted within 10 days of the incident occurring.
- **Over-seven-day injuries** must be reported using the online report form within 15 days of the injury occurring. There is no requirement for immediate notification.
- **Reportable diseases** must be reported using the online report upon written confirmation of diagnosis by a medical practitioner.
- **All reportable events** are to be reported online, using the appropriate F2508 form (of which there are various versions).

Type of Event	Immediate Notification Required?	Written Report Timescale
Fatality	Yes - online or by phone	10 days
Specified injury	Yes - online or by phone	10 days
Dangerous occurrence	Yes - online	10 days
Over-seven-day injury*	No	15 days
Reportable disease	No	'Forthwith' After diagnosis by a doctor
*Over-seven-day injuries must be notified within 15 days of the incident. Over-three-day injuries are recordable but not reportable.		

MORE...

Further information on **RIDDOR** can be found at:

www.hse.gov.uk/riddor/report.htm

Record-Keeping

Records of reportable events must be kept by the responsible person for a **minimum of three years**. This is usually done by printing out and keeping a copy of the online **RIDDOR** report form.

RIDDOR also require that a record of all 'over-three-day injuries' is kept for a minimum of three years.

An 'over-three-day injury' is a work-related accident that causes an injury that results in a worker being away from work or incapacitated for more than three consecutive days (not counting the day of the accident but including any weekends or other rest days). Over-three-day injuries **do not** have to be reported to the authorities unless the incapacitation period goes on to exceed seven days. Normally, the standard injury report form that must be kept for all work-related injuries is sufficient to meet the over-three-day injury recording requirement.

An example of Form F2508IE Report of an Injury follows. Make a note of the main headings and the basic information that has to be submitted.

Health and Safety Executive

Close window

Report of an injury

Note: this is a preview of your form and does NOT represent the submitted details of your notification, which will include the Notification number for reference

About you and your organisation

Notifier name			
Job title			
Organisation name			
Address			
Phone no		Fax Number	
Email Address			

Where did the incident happen

About the incident

Incident Date		Incident Time	
In which local authority did the incident occur (Country, Geographical Area and Local Authority)?			
..			
In which department or where on the premises did the incident happen?			
What type of work was being carried out (generally the main business activity of the site)?			
- -			

About the kind of accident

Kind of accident	
Work process involved	
Main factor involved	
What happened	

About the injured person

Injured persons name			
Injured persons address			
Phone no		What was their occupation or job title?	
Gender		Age	
Work Status			

About the injured person's injuries

Severity of the injury			
Injuries		Part of the body affected	

F2508IE Report of an Injury

Source: HSE (https://extranet.hse.gov.uk/lfserver/external/F2508IE)

STUDY QUESTIONS

15. Who is usually responsible for initially reporting accidents and other safety-related incidents?
16. What is the purpose of analysing information about accidents?
17. To whom do the results of accident investigations have to be communicated?
18. What are the six main categories of reportable incidents under **RIDDOR**?
19. Which events have to be reported straightaway under **RIDDOR**, and how?
20. For how long should records of reportable incidents be kept?
21. Which occupational diseases must be reported?
22. Apart from the accident documentation about the particular incident, what other information may be relevant to a claim for damages in respect of injuries suffered as a result of a guard malfunctioning on a drill?
23. What actions should be taken following a serious injury at work?

(Suggested Answers are at the end.)

Summary

This element has introduced you to managing change and procedures.

In particular, this element has:

- Outlined how change in the workplace can be safely managed.
- Described the number of factors that influence the impact of change such as communication, risk assessment, competency, welfare provision.
- Explained that change needs to be constantly reviewed to ensure that the construction plan is being kept on track.
- Defined SSOWs as formal procedures based on a systematic examination of work equipment and processes, to identify the hazards involved and define safe methods of working to eliminate or minimise risk.
- Identified the role of both competent persons and employees in developing and documenting these safe systems.
- Outlined the process of task analysis, where a task is broken down into a series of steps so that hazards can be identified and risk controlled at each step. This safe working method must then be implemented and monitored.
- Outlined permit-to-work systems as a way of formalising the control of high-risk work activities.
- Explained the Issue, Receipt, Clearance and Cancellation sections typical of a permit to work and outlined the use of permits in the control of hot work, work on non-live (isolated) electrical systems, entry into confined spaces, work at height and some forms of machinery maintenance work.
- Described the emergency procedures that an organisation should develop to deal with foreseeable incidents.
- Outlined how employers must provide and notify staff of its first-aid facilities, equipment and personnel.
- Explained the fire procedures that must be in place so that employees are aware of what actions to take in the event of fire such as means of escape, emergency evacuation procedures, the role of fire marshals/wardens, fire drills and provisions for people with disabilities.
- Outlined suitable emergency arrangements that must be provided when working near water.
- Outlined that emergency procedures should be continually reviewed and updated during a build to reflect any changes and should consider the use of Temporary Accommodation Units (TAUs).
- Described a basic investigation procedure:
  - Gather factual information about the event.
  - Analyse that information to draw conclusions about the immediate and root causes.
  - Identify suitable control measures.
  - Plan remedial actions.
- Outlined the arrangements that should be made for the internal reporting of all work-related incidents and the records of work-related injuries that should be kept.
- Described the requirements of **RIDDOR** and the types of incident that have to be reported to the enforcing authorities, such as fatalities, specified injuries, dangerous occurrences, occupational diseases, immediate hospitalisation of non-workers and over-seven-day injuries.

Exam Skills

Question

Scenario

Following your meeting with the site manager and concerns about 'out-of-date' risk assessments, you decide to check if some of the control measures identified in those risk assessments have been put into practice. One of the control measures identified was 'emergency procedures are practised regularly' on site. This action was assigned to the site foreman. However, there is no evidence to suggest emergency situations have been practised. None of the workers you have spoken to can remember the last time they did a 'fire drill'. You decide to talk to the foreman giving reasons why workplace emergency procedures should be practised. The foreman is often accused of being argumentative and difficult to motivate into taking action.

Task: Emergency Procedures

What reasons would you give to encourage emergency procedures to be practised?

(Total: 10 marks)

Approaching the Question

Now think about the steps you would take to answer this question:

Step 1 The first step is to **read the scenario carefully**. There is evidence that risk assessments are not suitable and assigned actions for control measures have not been completed. You are going to speak to the foreman and the conversation needs to be convincing enough to encourage action to be taken. You have been advised that convincing the foreman to take action may be difficult.

Step 2 Now look at the **task** - you need to provide reasons why emergency procedures should be practised.

Step 3 Next, consider the **marks** available. In this task, there are 10 marks available so it is expected that around 10 or 11 different reasons for practising emergency procedures need to be provided.

Step 4 **Read the scenario and task again** to make sure you understand them and have a clear understanding of why we should practise emergency procedures. (Re-read your study text if you need to.)

Step 5 The next stage is to **develop a plan** - there are various ways to do this. Remind yourself, first of all, that you need to be thinking about 'reasons to practise emergency procedures'. So, the answer plan will take the form of a bullet-pointed list that you need to develop into your answer.

Suggested Answer Outline

Reasons:

- Legal requirement.
- Internal policy requirement.
- Insurance company requirement.
- Vulnerable persons.
- Reduce likelihood of panic.

- Test communications.
- Those designated to assist can practise skills.
- Management commitment.
- Identify weaknesses.
- Familiarise employees with emergency procedures.
- Establish the alarm can be heard.

Now have a go at the question yourself.

Example of How the Question Could be Answered

Reasons why we must practise emergency procedures include:

1. *It is a legal requirement to practise certain procedures (e.g. fire drills). If the site had a fire then it's likely the cause would be investigated. If the investigation finds emergency procedures have not been practised then it's possible a fine may be imposed.*
2. *An organisation's own policy and procedures may stipulate that emergency procedures must be practised.*
3. *It's quite likely, because of the nature of business for our organisation, that our insurance company will require emergency procedures to be practised.*
4. *Practising emergency procedures allows for a speedy and correct response where vulnerable persons are involved. We have a clear duty of care for vulnerable workers, e.g. young workers or lone workers. It would be inexcusable if, through our inaction, harm was caused.*
5. *It will reduce the likelihood of panic if a real emergency situation arises. It's difficult to predict how people will behave in an emergency. But if workers panic, the likelihood of someone being harmed will increase.*
6. *Practice may reveal shortfalls in communication, either radio communication or verbal communication. We could simulate one exit being unavailable due to the emergency. We could then observe people's behaviour and change any shortfalls in our procedures.*
7. *It allows those with special duties to also practise their duties. The organisation has appointed fire marshals to assist in the event of an emergency due to fire.*
8. *Practising emergency procedures demonstrates management's commitment. Workers' morale can be improved if they believe the organisation cares for them as people – not just workers.*
9. *It may highlight problems and weaknesses in the emergency response that can then be addressed. We can establish, under safe conditions, that the emergency assembly point is well identified and everyone knows where it is.*
10. *Sounding the alarm in a premise will ascertain that the alarm can be heard in all parts of the premises. It certainly sounds loud enough when you are underneath one of the sounders, but we need to establish it's loud enough when all the equipment is operating also.*

Reasons for Poor Marks Achieved by Exam Candidates

- Providing a short answer that only addresses one or two points, probably due to not researching the answer or not paying attention to the number of marks being awarded. This approach will only achieve low marks.
- Failing to elaborate on points in the answer. Providing an answer similar or identical to the suggested answer outline may result in failing to achieve full marks.
- Wasting time by giving two reasons (usually legal and familiarising employees), then going on to discuss the duties fire marshals may have. This is far too narrow a view and cannot attract many marks.

Element 4

Excavation

Learning Objectives

Once you've studied this element, you should be able to:

1. Explain the hazards and risk assessment of excavation work.
2. Explain the control measures for excavation work.
3. Explain the hazards and risks associated with confined space working and safe working.

Contents

Excavation Work Hazards and Assessment

IN THIS SECTION...

- The hazards of excavations include: striking buried services; people, objects or materials falling in; collapsing of sides; collapsing of adjacent structures; flooding; using cofferdams and caissons; contaminated ground; toxic and asphyxiating atmospheres; and mechanical hazards.
- Consideration should be given to overhead hazards, including power lines.
- Risk assessment factors to consider are: depth, type of soil, type of work, use of mechanical equipment, proximity of roadways and structures, the presence of the public and weather conditions.

The Hazards of Work in and around Excavations

The following are hazards associated with excavation work.

Buried Services

Striking services such as high-voltage electricity cables, gas pipes, mains water or other buried services (e.g. telephone and cable TV lines) can lead to electric shock, arcing, burns, fire, gas explosion or rapid flooding of the excavation, as well as major business disruption to service users in the area.

A shallow excavation revealing a collection of buried services

Falls of People/Equipment/Material into Excavation

Falls can happen:

- Because of an unfenced edge.
- While people are climbing in or out from ladders or other access equipment.
- When tools or materials (bricks, timber, etc.) fall into an excavation onto persons, from an unprotected edge.
- When vehicles drive too close to the side of an excavation, collapsing the sides or tipping in.
- When spoil (loose soil) or stacked, loose sand is piled too close to the sides of an excavation.
- When adjacent structures (e.g. wall or scaffold) are undermined by an excavation and collapse in.

Collapse of Sides

Collapse occurs when the unsupported sides of an excavation slip and cave in (often due to poor support systems of the excavation sides). Severe crush injuries can result from even relatively small collapses because soil is very heavy, especially when wet. Workers buried or trapped in soil can asphyxiate in minutes, and do not have to be completely buried for this to occur; being buried up to the chest can lock the rib cage and have the same effect.

Collapse of Adjacent Structures

Digging too close to, or under, the foundations which support nearby buildings or structures may undermine their support and cause collapse of the building or structure into the excavation. This would be even more dangerous if the excavation itself were also to collapse.

Water Ingress

Flooding can occur from:

- Surface water, during heavy rain or snow.
- Groundwater (a high-area water table), nearby rivers, streams and watercourses (especially if breached).
- A burst water main caused by the excavation activities.

Use of Cofferdams and Caissons

These are structures (usually watertight enclosures) which are pumped dry to allow work to be carried out inside them below the waterline on bridge building, etc.

They must be suitably designed and constructed to prevent the ingress of water (or other materials), or appropriately equipped to pump out water, and provide shelter and escape should water or materials enter it. The cofferdam or caisson must be inspected by a competent person at the start of every shift or after any event likely to have affected its strength or stability.

Cofferdams or caissons constructed to prevent the ingress of water

Contaminated Ground

Contaminated ground can:

- Be found on sites that previously housed chemical works or storage areas.
- Contain methane or hydrogen sulphide gas (both from microbial decay).

Contaminants can be varied in range and include the following:

Industry	Possible Contaminant
Petrochemicals	Hydrocarbons, benzene, phenol, acids, alkalis
Steel/iron works	Iron, copper, zinc, asbestos
Gasworks/power stations	Coal, sulphur, phenol, asbestos, cyanides
Pits/quarries	Leachates, copper, zinc, lead, methane
Tanneries	Anthrax
Miscellaneous industries	Polychlorinated biphenyls, sulphates, metals, micro-organisms

Toxic and Asphyxiating Atmospheres

Toxic and asphyxiating atmosphere are created from the industries mentioned in the table above and from gases used on site. The hazards are that:

- Heavier than air, gas such as Liquefied Petroleum Gas (LPG) and carbon dioxide, can infiltrate an excavation.
- The combustion gases from nearby construction equipment, such as diesel generators and motor vehicles, can seep into excavations with the same effect.

Mechanical Hazards

Mainly from the use of plant and equipment around or in the excavation:

- Vibration from plant operation may cause collapse.
- Excavating machinery itself may create hazards, such as striking persons in or around excavations.

Overhead Hazards Including Power Lines

Work beneath overhead services and power lines in particular should be avoided.

Before excavation work starts, all overhead services should be identified and any diversions or disconnections ensured before excavation work begins. Service providers should be contacted to obtain accurate plans of supplies.

Three situations arise in construction work at overhead power lines:

- No scheduled excavation work or passage of plant to take place under the lines.
- Excavation plant and equipment will pass beneath the power lines.
- Excavation work will take place beneath the power lines.

These situations, and appropriate controls, are discussed in detail in Element 10 'Electricity'.

Risk Assessment

Factors to Consider

Risk assessments should be carried out in accordance with the **Management of Health and Safety at Work Regulations 1999 (MHSWR)**, with due consideration to the excavation under the **Construction (Design and Management) (CDM) Regulations 2015**.

TOPIC FOCUS

Risk assessment - factors to consider:

- Depth of the excavation.
- Soil type - this may vary from fine sand which can flow easily, to heavy clay which is much more cohesive.
- Three broad classes of ground exist:
  - Non-cohesive ground or light soil, e.g. sand or gravel, whose natural angle of repose when dry is usually 45° or less.
  - Cohesive ground or heavy soil, e.g. stiff clay, whose natural angle of repose is about 60°.
  - Rock, whose natural angle of repose varies from about 80° for loosely bonded or light rock, to 90° for tightly bonded heavy rock. Rock may, however, have steeply sloping clay planes which may fail, with resulting collapse.
- Type of work involved, e.g. at the side of a road; in a housing development; laying pipes/cables; trenches; pits.
- Use of mechanical equipment - the types being used.
- Proximity of the excavation to roadways, watercourses, structures, schools and hospitals.
- Presence of the public/children.
- Weather.

STUDY QUESTIONS

1. What hazards are associated with work in and around excavations?
2. What factors should be taken into account with regard to risk assessments for excavations?

(Suggested Answers are at the end.)

Control Measures for Excavation Work

IN THIS SECTION...

- Precautions must be taken to prevent persons falling into, or being injured while working in, excavations. General precautions include:
  - Identifying buried services and using safe digging methods to avoid contact with them.
  - Supporting the sides of excavations to prevent collapse, and providing workers with suitable access and egress, and crossing points to pass over excavations.
  - Barriers, lights and signs used to demarcate danger areas.
  - Spoil to be removed to and stored at a safe distance from the excavation to prevent it collapsing back in.
  - De-watering methods used to remove water from excavations.
  - Vehicles and materials on site to have special precautions applied to prevent them falling into or collapsing the excavation.
  - Workers in excavations to wear items of Personal Protective Equipment (PPE) such as hard hats and safety footwear.
- When working in the vicinity of contaminated ground, testing is essential and extra welfare facilities should be provided to accommodate workers (separate from normal site facilities). Health surveillance may be appropriate.
- Excavation supports are to be inspected before each shift and after any event that could affect the integrity of the excavation, and reports to be made and kept.

Controls

In common with other construction activities, control of the risks involved in excavation is based on effective management. The **CDM Regulations 2015** apply in this respect, and excavations must be carried out under the supervision of a competent person.

The Health and Safety Executive (HSE) guide *Avoiding danger from underground services* (HSG47) provides further detailed reading on this subject.

Identification/Detection and Marking of Buried Services

The location and configuration of underground services should be identified prior to work commencing. It may well be possible to avoid cable routes at the planning stage of work. Before work starts, the following action should be taken:

- Check any available plans.
- Contact local service providers and owners, such as electricity, gas, water, telecommunications or TV companies.
- Survey the site and surrounding areas to identify indicators of the existence of cables, etc., e.g. streetlights or junction boxes.
- Use cable locators with trained operators. Plastic and non-metallic underground services cannot be identified by conventional locators, but could be identified by the use of metallic tracer wire laid with the pipe or by using a signal transmitter inserted and pushed along the pipe itself.

The positions of known services should be marked on plans and also on the ground itself. All employees must receive adequate information and instruction about the nature of the risks.

Where appropriate, arrangements must be made with the service providers to isolate the cables/pipes and ensure that it is safe to work in the vicinity of them.

Safe Digging Methods

Safe digging methods include:

- Using locators to determine the position and route of pipes or cables (frequently using them during the course of the work).
- Keeping a careful watch for evidence of pipes or cables. Remember that plastic pipes cannot be detected by normal locating equipment.
- If contact is made with any unidentified service pipe or cable, stopping work until it is safe to proceed.
- Regarding all buried cables as live until disconnected and proven - pot-ended cables cannot be assumed to be dead or disused.
- Not using excavators and power tools within 0.5m of the indicated line of a cable/pipe.
- Using hand digging when nearing the assumed line of the cable/pipe.
- Using spades and shovels (preferably with curved edges to aid scooping) rather than other tools, e.g. forks and picks.
- Reporting any damage to the appropriate services and keeping personnel clear until it is repaired.
- Having an emergency plan to deal with such damage to pipes or cables.
- Supporting exposed cables and pipes and protecting against damage by backfilling. They should never be used as hand- and foot-holds.

Use of a cable detector
(Source: HSG47 *Avoiding Danger from Underground Services*, HSE, 2014 (www.hse.gov.uk/pubns/priced/hsg47.pdf))

Methods of Supporting Excavations

Excavation supports will prevent the collapse of the side walls of the excavation and allow work to continue uninterrupted inside the workings. The type of support structure used will vary, depending on the:

- Type of ground being excavated.
- Length of time the excavation will be open and in use.
- Type of work being carried out.
- Groundwater conditions and potential for flooding.
- Depth of the excavation.
- Number of people in the excavation.

Here are some examples of types of **excavation supports**.

Steel Sheets

These are laid across the faces of the excavation walls, secured together by clamps and braced by expandable steel struts or wooden beams. The steel sheets should be toed in at the bottom and rise above the tops of the trench sides.

Shoring the sides of an excavation by 'close sheeting'

Support Boxes

Support boxes (also known as trench boxes) are ready-constructed units which can be easily and quickly installed into an excavation using an excavator or similar machine, to provide strong, reliable shoring for most ground conditions while giving full protection for operatives. They can be moved along as the work progresses.

Where necessary, both types should have edge protection and a means of access via a protected area.

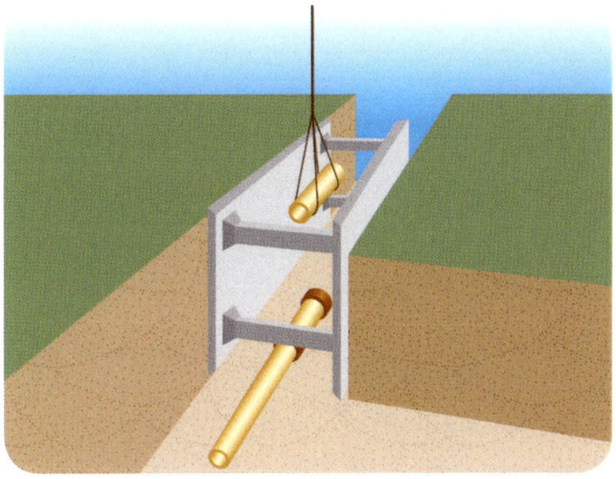

A trench box placed inside a trench creates a protected area for workers

Means of Access

Ladders provide the main method for access to and egress from an excavation. They must be suitably secured to prevent undue movement and extend above the excavation to give the necessary height required for a safe handhold (at least 1 metre or five rungs of the ladder).

There must be adequate means of escape in an emergency: one ladder every 15m is an average to work to (more may be required depending on the number of workers and the potential risk, e.g. where there might be a possibility of flooding from a rising water table).

Ladders must be kept in good condition, be fit for the purpose and of adequate strength. Climbing in and out of the excavation using other means should be prohibited.

Crossing Points

Excavations should only be crossed at designated points. The crossing point should be of sound construction and suitable to support all the types of vehicle and equipment likely to use it. Gangways across excavations should have guardrails and toe boards.

Crossing point

Barriers, Lighting and Signs

Barriers

Barriers should consist of guardrails (as for a scaffold work platform) to prevent people falling in, and toe boards to prevent objects being kicked down into the excavation. In addition:

- To prevent vehicles from falling in, logs or concrete blocks are laid some distance from the edge to act as a buffer.
- Excavations may need to be covered, particularly at night. Such covers need to be capable of bearing a person's weight and be held securely in place.
- Fencing or hoarding may be required to protect members of the public as well as construction employees.

Lighting

Appropriate lighting should be installed to ensure that there is an adequate level of illumination, without distracting shadows, to ensure the safety of work activities both within the excavation and on the surface. High-powered electric lights or those which operate from LPG will be required for general workplace illumination, and consideration may need to be given to smaller, personal lights for individual workers. These should be battery-operated to avoid the risks associated with trailing electrical leads.

Signs

Signs should be used to warn people of the excavation hazards, and any special precautions required. They should be placed in clearly visible spots at all potential access points.

When working on a roadway, the police or local authority need to be consulted about traffic lights or stop/go signs. There are rules for the placing of cones to warn motorists of the hazard.

Safe Storage of Spoil

If spoil (the ground dug from an excavation) is to be left in spoil heaps near the excavation, there must be space to stack it without interfering with other operations, leaving a clear area of 0.6m between the spoil and the edge of the excavation. The officially recognised depth of the excavation depends on **the distance of the spoil heap from its edge**. For excavations up to 6m deep, the following rules apply in the UK:

- If the spoil heap is within 1.5m of the edge of the excavation, the depth of the excavation should be measured **from the top of the spoil heap**.
- If the spoil heap is more than 1.5m from the edge (but less than the actual depth of the excavation), the official depth should be taken as the actual depth plus half the height of the spoil heap.

De-Watering Methods

Water can fall as rain, snow and sleet directly into excavations, especially if they are uncovered. The run-off following these downpours can also run in. Water can enter through the sides depending on the height of the surrounding water table, and close proximity to watercourses (rivers, streams, lakes, etc.) can cause excess water.

Accidental damage to nearby tanks, water mains or other supply pipework in the vicinity of the excavation can also lead to an ingress of water. All exposed pipes should be identified and supported.

To protect the stability of the sides of the excavation, drainage channels can be cut around the excavation and the water can be channelled away to sump points where it can be pumped away. The water table is then lowered below the level of the excavation. In extreme cases, watercourses might be redirected.

Excavation sides can be strengthened by using higher shoring and sandbagging the outside. However, the continued flow of water from the area surrounding the excavation may cause settlement problems.

The disposal of any water should be discussed with the appropriate environmental agency.

Well Points

Well pointing or deep wells can be considered where sandy or silty soil exists. Ground freezing or providing an impermeable barrier by injecting cement, bentonite or a chemical is an option, but can be expensive and cause disruption to the ground.

Sump Points

Creating **sump points** is the most common and economical method of de-watering as gravity is the main playing force. Sump is created in the excavated area into which the surrounding water converges and accumulates facilitating easy discharge of water through robust solid handling pumps. Its application is, however, confined to the areas where soil is either gravelly or sandy.

Positioning and Routeing of Vehicles, Plant and Equipment

To prevent vehicles and objects falling into excavations, and collapse of sides, the following precautions should be observed:

- Machinery and vehicles should only be operated in specially designated areas strong enough to withstand the effects of vibration without causing a collapse. Specially built routes to allow access to the floor of the site for vehicles and plant may have to be constructed.
- Where vehicles have to pass close to excavations, barriers and signs should be in place to keep them a safe distance away.
- Where vehicles have to approach the edges of the excavation (e.g. to receive or tip materials), there should be stop blocks to prevent them overrunning.

Personal Protective Equipment

Hard hats and safety footwear are a likely requirement at all times. The need for other PPE will be determined by the nature of the work carried out in the excavation, for example:

- Fumes and dust may require the use of masks and respirators.
- Excessive noise levels will require the wearing of hearing protection.
- Where welding work is carried out, face shields and protective clothing will be necessary.
- Breathing apparatus and safety harnesses may be required for working in tunnels and shafts.

Particular Requirements for Contaminated Ground

Soil Testing

Soil testing to show the presence of any contaminants must be carried out by competent persons from an accredited laboratory.

Soil testing

Once the nature of contaminants is established, measures can be put in place to limit the risk that might be present, e.g. PPE, hygiene facilities, Safe Systems of Work (SSOWs) and medical surveillance.

Great care must be taken as:

- Digging will uncover buried materials or contaminants within the ground which are hazardous to health.
- The contaminants themselves can change over time due to bacterial or chemical action, which may also alter their properties. This may be the result of the decomposition of organic matter or from the dumping or spillage of hazardous substances.
- Certain contaminants are subject to specific legislation, namely asbestos, lead, anthrax, radioactive materials and buried explosives, and well-established procedures need to be followed to deal with each of these situations.
- There may be radioactive hazards from the ground itself (probably as a result of previous occupancy).
- Toxic and asphyxiating atmospheres may be present, for example:
  - Flammable gases such as methane (marsh gas) and carbon monoxide from plant and machinery used in connection with excavation, including pumps involved in de-watering operations.
  - Toxic gases such as hydrogen sulphide.
  - Chemicals and metal compounds, either in containers or within the soil.

Welfare Facilities

Where work involves contaminated ground, the welfare requirements include:

- Separate facilities (e.g. toilets, washrooms and showers, rest and eating facilities) from the main facilities for all work involving contaminated sites.
- Specialist facilities may be required where work involves asbestos, lead or radioactive materials.
- Separate changing rooms and clothing accommodation may be required to prevent cross-contamination.
- Adequate heating, lighting and ventilation in these facilities.

Health Surveillance

Medical practitioners can screen the workforce involved (e.g. blood/urine testing) and provide continued monitoring to ensure that no serious medical conditions arise. The effect of screening can also alleviate any anxiety caused to either the workforce or the public as to whether the measures in place are adequate.

Inspection Requirements for Excavations and Excavation Support Systems

Regulation 22(4) of the **CDM Regulations 2015** deals with the frequency of inspections.

No person is allowed to work in an excavation (other than to examine it) until it has been examined by a competent person.

The inspection of excavations and their supports occurs:

- At the start of each shift.
- After any event likely to have affected the strength or stability of the excavation (e.g. flooding).
- After any material unintentionally falls or is dislodged.

An **inspection report** must be made and kept (CDM Regulation 24). This report must be:

- Completed before the end of the shift in which the excavation was inspected.
- Copied and this copy given to the person in charge of the worksite within 24 hours of completing the inspection.
- Retained at the site until the excavation work is completed and, after that, for three months.

Though excavations must be routinely inspected prior to each shift, no more than one written report is needed in seven days unless the strength or stability of the excavation is affected, e.g. after accidental falls of material.

TOPIC FOCUS

An **excavation inspection report** must contain:

- The date, time, location and a description of the workplace inspected.
- The name and position of the person inspecting/making the report.
- The name of the person on whose behalf the inspection was carried out (i.e. the person in charge of the worksite).
- Details of any matters identified that could give rise to health and safety risks, and any actions taken to control them.
- Details of any further actions considered necessary.

When work in the excavation is completed, support materials (timber, sheeting, etc.) should be safely removed by experienced workers and a competent person should inspect the site to ensure that all dangerous materials and equipment have been removed. Water may have to be pumped out of the excavation before it is filled.

Filling should use only appropriate materials and be conducted in a controlled manner under the direction of a competent person.

STUDY QUESTIONS

3. What controls may be necessary during excavation work?
4. What particular hazards may arise during the excavation of contaminated ground?
5. What inspection requirements are there for excavations?

(Suggested Answers are at the end.)

Safe Working in Confined Spaces

IN THIS SECTION...

- A confined space is any enclosed space with a specified risk of injury associated with it, including trenches, sewers, chambers, tanks, pits and cellars.
- Hazards include toxic, explosive and oxygen-deficient atmospheres; heat; water; free-flowing solids; and restricted space.
- Assessing risk and precautions ensures a safe system of work. Precautions include avoidance, planning, training, testing the atmosphere, providing means of access, appropriate personal protective equipment, and monitoring and emergency arrangements.
- A permit to work may be needed but is not required if the risk can be controlled, the working system is simple and other activities do not affect safe working.

Types of Confined Spaces and Why They Are Dangerous

DEFINITION

CONFINED SPACE

Any enclosed space where there is a reasonably foreseeable specified risk of any serious injury associated with it.

Typical confined spaces found in construction work include trenches, sewers, manholes, tunnels, excavations, chambers, tanks, pits, cellars and unventilated rooms. Cofferdams and caissons, by their enclosed nature, may also be included.

Other places can become confined spaces:

- Due to the type of work undertaken, e.g. spray painting a room.
- As a result of a change in conditions inside the space, e.g. overheating in a plant-room or electricity sub-station.
- Due to a change in confinement or the degree of enclosure, e.g. inside an item of moving plant or machinery (such as a compactor or baling machine).

TOPIC FOCUS

'**Specified risks**' associated with work in confined spaces are risks of:

- Serious injury arising from fire or explosion.
- Loss of consciousness arising from an increase in body temperature.
- Loss of consciousness arising from asphyxiation due to gas, fumes, vapour or lack of oxygen.
- Drowning as a result of an increase in the level of liquids.
- Inability to reach a respirable environment due to entrapment by a free-flowing solid, e.g. flour, grain, sand, sugar, etc.

The **Confined Spaces Regulations 1997** lay down the legal requirements for work in confined spaces. These Regulations contain the following key duties:

- Avoid entry to confined spaces, e.g. by doing the work from outside.
- If entry to a confined space is unavoidable, follow an SSOW and put in place adequate emergency arrangements before the work starts.

Due to the nature of the work itself, i.e. where there are risks from machinery, electricity or hazardous substances, other legislation may be involved as discussed in Elements 9, 10 and 12.

The Main Hazards Associated with Working within a Confined Space

Some of the conditions below may be present in the confined space, some may arise from the work being carried out, or because of ineffective isolation of plant nearby, e.g. leakage from a pipe connected to the confined space.

The enclosure and working space may increase other dangers arising through the work being carried out, such as:

- Machinery being used may present addition hazards, such as:
  - Dust, e.g. from a portable grinder.
  - Electric shock.
- Gas, fumes or vapour can arise from welding, or by use of volatile and often flammable solvents or adhesives.

Atmospheres

Toxic Atmospheres

Poisonous gases, fumes and vapours can build up in sewers

Poisonous gas, fumes or vapour can:

- Build up in sewers, manholes and pits.
- Enter tanks or vessels from connecting pipes.
- Leak into trenches and pits in contaminated land, e.g. old refuse tips and old gas works.

The following toxic gases may be present:

- Hydrogen sulphide from sewers or decaying material.
- Carbon monoxide from internal combustion engines (petrol/diesel) or the incomplete combustion of LPG.
- Carbon dioxide:
  - From fermentation processes.
  - Naturally from the rocks/soil.
  - From combustion processes.
- Certain fumes/vapours from chemicals, e.g. ammonia, chlorine or petrol/solvents.

Explosive or Flammable Atmospheres

Certain substances, even in small concentrations, can pose a hazard and produce either a fire or an explosion depending on their explosive range in the atmosphere. Examples of gases include LPG, acetylene, propane, butane, methane, hydrogen and hydrogen sulphide. Examples of liquids that have flammable vapours include acetone, toluene, alcohol, white spirit, thinners and solvents.

Confined spaces that need to be entered for cleaning or maintenance may have been used to store hazardous substances, e.g. toxic, flammable or explosive liquid or dust.

Oxygen-Deficient Atmospheres

Oxygen-deficient atmospheres can occur:

- Through displacement (the addition of a gas or vapour to the space displaces the oxygen), for example:
  - Purge gas.
  - Pipe freezing.
  - Gas leaking from elsewhere.
- Through consumption (oxygen in the atmosphere is depleted by a chemical reaction or biological process occurring in the space), e.g. oxidation, rusting, bacterial growth.
- Through welding operations.
- By people working.
- By any process of combustion.
- Where there is a reaction between some soils and the oxygen in the atmosphere.
- Following the action of groundwater on chalk and limestone - this can produce carbon dioxide and displace normal air.

Heat

Hot conditions can lead to a dangerous increase in body temperature and may result in:

- Heat stress, fatigue or exhaustion.
- Dehydration/loss of fluids.
- Loss of consciousness.

Water

Work in confined spaces such as trenches, pits or sewers that contain standing water can lead to gastroenteritis, Weil's disease and hepatitis A or B. There is also a risk of water ingress and being drowned or swept away.

Free-Flowing Solids

Free-flowing solids are substances consisting of solid particles that, because of their small size and numerous quantities, have a flowing or running consistency. Examples include cement, sand, grain and coal dust.

A person can drown or asphyxiate if immersed in a free-flowing solid. The solid acts as a liquid and so presents the same hazards as a liquid. If breathed in, the solid could clog the respiratory tract and enter the lungs, depriving the person of oxygen.

Restricted Space

This is where a confined space imposes exceptional constraints as a result of its physical layout. It may be further restricted by the equipment it contains, and the tools and PPE used by the person entering.

Liquids or solids can suddenly fill the space, or release gases into it, when disturbed. Free-flowing solids, such as cement, can also partially solidify or 'bridge' in silos, causing blockages which can collapse unexpectedly. If access to the space is through a restricted entrance, e.g. a manhole, escape or rescue in an emergency will be more difficult.

What Should be Considered when Assessing Risks from a Confined Space

The **MHSWR** require a risk assessment to be carried out to identify the hazards present, assess the risks and determine what precautions to take. The **Confined Spaces Regulations 1997** apply, and the assessment will help to identify the precautions to be planned and included in the SSOWs required. Appropriate health surveillance must be offered where required.

TOPIC FOCUS

Confined space risk assessment will consider:

- Whether confined space entry can be avoided.
- Work needing to be done, e.g. routine or breakdown.
- Methods of working - tools and equipment to use.
- Entry with or without breathing apparatus.
- Hazard identification in the plant.
- Hazards from neighbouring plant/vessels.
- Suitability of those carrying out the work.
- Steps necessary to make the job safe, i.e. an SSOW.
- Arrangements for emergencies and rescue facilities.

The Precautions to be Included in a Safe System of Work for Confined Spaces

Avoidance Where Possible

Persons should not enter a confined space to carry out work unless it is not reasonably practicable to do it any other way. Better planning of work or using a different approach can reduce the need for confined space working, for example:

- Modify the confined space itself so that entry is not necessary.
- Do the work from outside, e.g. blockages can be cleared in underground chambers by use of remotely operated rotating flail devices, vibrators or air purgers.
- Do inspection, sampling and cleaning operations from outside the space using appropriate equipment and tools.
- Use remote cameras for internal inspection of vessels.

An SSOW is always required to enter, carry out work in, and leave a confined space (other than in an emergency).

Avoid entry where possible

Planning

Preliminary planning of confined space entry is important to establish whether entry:

- Can be avoided or an alternative method used.
- Is required with or without breathing apparatus.

Proper planning will also help to establish effective lines of authority and communication so that the procedures put in place are clear and free from any misunderstanding or conflict of interest. The precautions necessary for each entry will vary with the nature, size, location, risks involved and number of people working, however, the basis of the SSOW will depend on the risk assessment carried out.

Training and Use of Competent Persons

All persons involved with supervising or carrying out confined space work must have adequate training, which should be repeated regularly.

The training should involve the permit to work, respiratory equipment, gas-testing equipment, rescue procedures, first-aid treatment including artificial respiration, evacuation and emergency procedures, fire-fighting and communication procedures. No person should be allowed to enter a confined space unless they are trained and competent to do so and records should be kept of all types of training carried out.

Atmospheric Testing

It is necessary to check that the atmosphere within a confined space contains sufficient oxygen, is free from both toxic and flammable vapours, and is fit to breathe. Testing should be carried out by a competent person using a suitable gas detector which is correctly calibrated. Breathing apparatus should be worn to enter a confined space for testing. Gas testing involves testing the atmosphere at all locations, drain points, instrument bridles, orifices, and at a high and low level, as some denser gases do lie in low positions, e.g. hydrogen sulphide.

Where the risk assessment indicates that conditions may change, or as a further precaution, continuous monitoring of the air may be necessary.

TOPIC FOCUS

Atmospheric testing should detect:

- A flammable gas, vapour or mist in excess of 10% of its Lower Explosive Limit (LEL).
- An atmospheric oxygen concentration below 19.5% or above 22%.
- An atmospheric concentration of any substance for which a Workplace Exposure Limit (WEL) is published in EH40 by the HSE and which could result in employee exposure in excess of the permissible limits.
- An airborne combustible dust at a concentration that obscures vision at a distance of 5 feet (1.52m) or less.
- Any atmospheric condition recognised as immediately dangerous to life or health.

Flammable

Explosive

Testing the atmosphere in confined spaces is essential. It is particularly important to be aware of any flammable or explosive substance that may be/have been present

Means of Access

Safe access/egress must be provided. No one may enter or remain in a confined space unless the following requirements are met:

- They are wearing approved breathing apparatus if dangerous fumes are present or in an oxygen-deficient atmosphere. The minimum space required in this case is 575mm. Smaller openings must be assessed for safe access and egress.
- Depending on the work being done, a number of access and egress points may be required.
- They have been authorised to enter by a responsible person, i.e. through a permit-to-work system.
- Where practicable, they are wearing a belt or harness with a rope securely attached.
- A person keeping watch (sentry) outside and capable of pulling them out is holding the free end of the rope.

A person may enter or work in a confined space without breathing apparatus, provided that:

- Effective steps have been taken to avoid ingress of dangerous fumes.
- Sludge or other deposits liable to give off dangerous fumes have been removed/cleaned.
- The space contains no other material liable to give off such fumes.
- The space has been adequately ventilated and tested for toxic, flammable and explosive gases/fumes.
- There is a supply of breathable air.
- The space has been certified by a responsible person as being safe for entry for a specified period without breathing apparatus.

Personal Protective Equipment

Appropriate PPE needs to be worn which:

- is fit for purpose;
- fits correctly; and
- is compatible with any other equipment worn.

In confined spaces this might include hard hats, coveralls, boots, wellingtons (sometimes called gumboots or galoshes), breathing apparatus sets, gloves and/or safety spectacles, and radios.

Monitoring Arrangements

Monitoring should be carried out by competent, trained personnel with individual detector tubes that are available for specific substances.

Gas monitoring can be carried out using appropriately calibrated portable gas detection equipment. The gas meters can be used to detect individual gases or to sample a range of gases.

Personal monitors may also be used by individuals working in confined spaces.

TOPIC FOCUS

Monitoring may be required:

- Before the first entry.
- At the start of each shift.
- At each entry.
- After each break.
- At a specified time limit, as set out in the permit.
- Continuously.

Emergency Arrangements

National or local regulations will require arrangements for emergency rescue to be in place when working in a confined space. The rescue plan must be communicated to all personnel involved before beginning work:

- Arrangements for raising the alarm and carrying out rescue operations must be in place.
- Rescue team members and equipment should be readily available for emergencies.
- Team members must be:
  - Properly trained in rescue techniques and use of all equipment.
  - Physically fit enough to carry out their task.
- Rescuers must be aware of and protected against the cause of the emergency.

Rescue equipment should consist of:

- Breathing apparatus sets.
- Rescue and resuscitation equipment - harnesses, lifting tripods and winch, stretchers, first-aid equipment.
- Fire-fighting equipment.
- Means of communication and summoning help:
  - A tug-rope is commonly used when personnel are not visible.
  - If radio is to be used, it must immediately be tested on entry.
- Lifelines.
- Oxygen - for resuscitation, not for 'sweetening' (increasing the oxygen content in the atmosphere).

Rescue teams must not enter without breathing apparatus. They must be thoroughly trained in rescue techniques, first aid and use of resuscitators.

An essential part of such a rescue plan might include:

- Shutting down adjacent plant before attempting emergency rescue.
- Communication with local emergency services (e.g. the fire service), remembering to pass on information about the particular dangers in the confined space.

When a Permit-to-Work for Confined Spaces Would Not be Required

An employer must consider a permit system for controlling confined space entry, using competent personnel to ensure formal checks are undertaken, and putting in place an SSOW before people are allowed to enter the confined space. A permit system is also a means of communication between site management, supervisors and those carrying out the hazardous work.

TOPIC FOCUS

Essential features of a permit-to-work system are:

- Clear identification of who may authorise particular jobs (and any limits to their authority).
- Who is responsible for specifying the necessary precautions (e.g. isolation, gas testing, emergency arrangements, etc.).
- The inclusion of contractors engaged to carry out work.
- Training and instruction in the issue and use of permits.
- Monitoring and auditing to ensure that the system works as intended.

However, according to HSE's *Safe work in confined spaces* (L101) - the Approved Code of Practice (ACoP) which supports the **Confined Spaces Regulations 1997** - a permit system is **unlikely to be needed** where, for example:

- the risk can be controlled easily;
- the system of work is very simple; and
- other work activities do not affect safe working in the confined space.

The decision not to use a permit should be taken by a competent person after consultation with specialists and having regard for the findings of the risk assessment.

STUDY QUESTIONS

6. (a) What is the definition of a confined space?

   (b) Where are confined spaces found on construction sites?

7. What hazards are associated with confined spaces?
8. What precautions are necessary for safe entry into a confined space?
9. What factors might be involved in a safe system of work for confined space entry?

(Suggested Answers are at the end.)

Summary

This element has introduced you to the hazards and precautions required to ensure safe work in excavations.

In particular, this element has:

- Described the hazards and risks associated with working in excavations, including buried services, falls into excavations, collapse of sides, collapse of adjacent structures, water ingress, use of cofferdams and caissons, contaminated ground, toxic and asphyxiating atmospheres, mechanical hazards and power lines.
- Considered risk assessment factors such as: depth, type of soil, type of work, use of mechanical equipment, proximity of roadways and structures, the presence of the public, and weather conditions.
- Described measures that will adequately control risks associated with excavations, such as marking of buried services; safe digging methods; methods of supporting excavations; means of access and crossing points; barriers, lighting and signs; safe storage of spoil, removal of water; safe routing and positioning of vehicles, plant and equipment; and appropriate PPE.
- Described particular requirements for ensuring safety when working in contaminated ground, and inspection requirements for excavations and excavation support systems.
- Defined what a confined space is and identified typical confined spaces found in construction activities.
- Described the hazards and risks associated with confined space working, including toxic, flammable and explosive atmospheres; heat, water and free-flowing solids; and work in restricted space.
- Outlined what to consider when conducting a risk assessment.
- Described the appropriate measures to control the risks associated with work in confined spaces, including avoidance where possible, planning, training and competence, atmospheric testing, means of access, appropriate PPE, and monitoring and emergency arrangements.
- Outlined circumstances when a permit to work may not be needed.

Exam Skills

Question

Scenario

The construction project is about to commence ground work activities on the site. The work will consist of creating excavations for new services coming on to the site. The site is a brownfield site with possible buried surfaces.

Task: Excavation Safety

You have been asked by the site manager to assist the foreman in creating a safe system of work for this phase of the project. Before meeting the foreman, you decide to prepare notes on the type of hazards likely to be encountered and measures required to mitigate the risks created.

(Total: 10 marks)

Approaching the Question

Now think about the steps you would take to answer this question:

Step 1 The first step is to **read the scenario carefully**. Note in this scenario you are required to identify hazards associated with excavation activities and what may be encountered while doing this work. In addition, you need to consider the type of control measures to mitigate the risk with each hazard.

Step 2 Now look at the **task** - the site has been used previously for other commercial or industrial activities, so there could be additional hazards that might affect the excavation work.

Step 3 Next, consider the **marks** available. In this task, there are 10 marks, so it is expected that around 5 hazards should be identified and an additional 5 control measures.

Step 4 **Read the scenario and task again** to make sure you understand them and have a clear understanding of how you would go about carrying out the excavation work. (Re-read your study text if you need to.)

Step 5 The next stage is to **develop a plan** - there are various ways to do this. One suggested way to do this would be to create a list of 5-6 hazards and next to each one appropriate control measures.

Suggested Answer Outline

Hazards associated with excavations:

- Premature collapse.
- Buried services.
- Falls of people and equipment.
- Water ingress.
- Contaminated ground.

Control measures:

- Ensure the excavation has supported sides to prevent collapse.
- Identify type and location of buried services (CAT scan) and develop digging techniques.
- Ensure the excavation has barriers and is illuminated if appropriate.
- Dewatering practices introduced.
- Examination to establish contaminate and provide additional welfare facilities.

Now have a go at the question yourself.

Example of How the Question Could be Answered

During excavation work, there are a number of hazards that could be encountered depending on the location where the work is likely to be carried out. The most common one is premature collapse of the excavation whilst it is being constructed or when work is being carried out inside it. To prevent this, supporting the sides of the excavations to prevent collapse, and providing workers with suitable access and egress, and crossing points to pass over excavations. The use of trench boxes can be used for this. Any excavation that lacks sufficient natural light will need to be illuminated to ensure safe working conditions.

Buried services on site could pose a serious hazard if encountered, especially gas mains or electrical services. To prevent contact identifying where they run via CAT scan, clearly marking them and then adopting suitable digging methods such as hand digging opposed to mechanical digging.

To prevent personnel and equipment falling into the excavation it is important to erect barriers around the excavation and post signage to warn of the dangers. The barriers need to be robust in construction to prevent mobile plant driving into the excavation.

Water ingress into excavations can occur after heavy rainfall or on ground with a high-water table. Introducing portable pumps to remove the water may be required and regular inspection of the excavation should also take place due to the weakening effect water may create.

Contaminated ground can increase the risk of ground work. In the first instance a survey needs to be conducted to establish the contaminates. Depending on the contamination, additional controls may need to be introduced such as PPE (respirators, overalls, etc.). Additional welfare facilities such showering and decontamination facilities may be required. Also, health surveillance regimes may have to be introduced subject to occupational health advice.

Reasons for Poor Marks Achieved by Exam Candidates

- Focussing on two or three points and going into too much detail.
- Providing a short answer that only addresses one or two points, probably due to not researching the answer or not paying attention to the number of marks being awarded. This approach will only achieve low marks.

Element 5

Demolition

Learning Objectives

Once you've studied this element, you should be able to:

1. Explain the meaning of the terms deconstruction, piecemeal demolition and deliberate controlled collapse.
2. Identify the main hazards of demolition and dismantling work.
3. Outline the control measures for demolition and dismantling work.
4. Identify the purpose and scope of a pre-demolition, deconstruction or refurbishment survey.

Contents

Demolition and Dismantling Hazards

IN THIS SECTION...

- Demolition and dismantling methods include deconstruction, piecemeal demolition, deliberate controlled collapse, pre-weakening, progressive demolition and wire rope pulling. It is important to select the appropriate method.
- There are many hazards and controls associated with demolition and dismantling work , including the premature collapse of buildings, falls and falling materials, plant and other vehicles, manual handling and exposure of workers to dust, fumes, noise and vibration.
- Other hazards include:
  - The existence of services, often buried, such as gas, water, electricity, etc.
  - The presence of hazardous substances.
  - Dilapidation.

Introduction to Demolition

Regulation 20 of the **Construction (Design and Management) (CDM) Regulations 2015** states:

> *"Demolition or dismantling*
>
> *(1) The demolition or dismantling of a structure must be planned and carried out in such a manner as to prevent danger or, where it is not practicable to prevent it, to reduce danger to as low a level as is reasonably practicable.*
>
> *(2) The arrangements for carrying out such demolition or dismantling must be recorded in writing before the demolition or dismantling work begins".*

The Meaning of Deconstruction

Deconstruction is a process of carefully disassembling a structure so that the material used during the initial construction can be reused. Deconstruction is generally carried out in the reverse order to construction.

The Meaning of Piecemeal Demolition

The piecemeal method of demolition uses hand-held tools or machines to gradually reduce the building to ground level. It is sometimes a preliminary to other methods. It can be begun or completed by the use of machines.

Techniques available for piecemeal demolition include:

- Pneumatic hammer.
- Hydraulic (pecker) or electric hammer.
- Tooth grapples.
- Ball and chain.
- Non-explosive blasting, using a liquid carbon dioxide cartridge which expands to a gas creating high pressure.
- Hydraulic bursters - these can be inserted into concrete to create bursting forces.

- Thermic lance method of melting concrete - the presence of reinforcement aids the process.
- Flame jet piercing - a melting process using a flame jet by burning kerosene in oxygen.
- Water jet - cement grouts can be cut with high water pressure.

For example, with a tall stack in an area where there are occupied buildings nearby, the task may be begun by slow, methodical hand dismantling until the structure has been reduced to about 10m before heavy dismantling plant can be used.

The Meaning of Deliberate Controlled Collapse

This is carried out using explosives or pre-weakening of the structure followed by remote mechanical demolition or pulling using a wire rope. The use of explosives requires a competent and fully qualified explosives engineer, and work must be properly documented and notified to various bodies (e.g. gas, electricity, rail, police, airports, local authorities). The process must be properly carried out using appropriate techniques regulated by the construction industry.

Selection of the Appropriate Method

Whether demolishing or dismantling a building or structure, it is vital that the pre-demolition survey identifies the correct method to be used.

Pre-Weakening

This is the removal of key structural members to cause complete collapse of the whole or part of the building or structure. Where controlled collapse demolition is adopted, the key structural members to be removed should be clearly indicated, together with the sequence of removal, in the method statement.

Certain problems can arise during this procedure, for example:

- The structure may collapse more than expected.
- Only partial collapse may occur leaving a dangerous structure to deal with.
- Considerable debris may be projected long distances, hitting unprotected people.
- The condition/state of the debris created may make it difficult to remove.

Progressive Demolition

This is the controlled removal of sections of the structure, while retaining the stability of the remainder and avoiding collapse of the whole or part of the building to be demolished. Progressive demolition (also known as 'top-down' demolition) should be considered for the majority of sites and is particularly useful in confined and restricted areas.

Progressive demolition: Remove part of a structure, leave the rest intact

Overturning by Wire Rope Pulling

This may also be used, except on brick or masonry buildings over 21m. The wire ropes are attached to winches or vehicles. The ropes should be attached prior to pre-weakening taking place and must be fit for purpose and not less than 38mm in circumference.

Equipment to be Used

Hand Tools

Demolition using hand tools can be efficient, requiring far less set-up time. Workers are required to work along the weaknesses of the section to be demolished, cutting along the mortar joints rather than through brick.

Where masonry/brickwork is involved, lifting operations may be necessary when cutting large structural members, and the debris needs to be disposed of using chutes or hoppers. Partial removal of floors may assist in the removal of debris, taking care to ensure the structural stability is maintained, and guardrails and toeboards are fitted for safety. Debris must be frequently disposed of and any debris that contains asbestos, radioactive materials or chemical contamination must be treated with all the necessary precautions. Services that may be needed must be identified and protected, and temporarily disconnected and reconnected by a competent person if required.

Certain precautions need to be in place:

- Access scaffolding should be progressively removed, ensuring no scaffold tubes project more than 3m above the demolition work.
- Scaffold towers, if in use, should not project above the work by more than one lift.
- The working platform should not be more than 6m above the highest row of ties (if in use).
- For roof work, crawling boards may be required which should span the purlins (horizontal timbers resting on the principal rafters, which support the common or subsidiary rafters or boards on which the roof is laid).
- Harnesses and/or edge protection are needed if recovery of materials is involved, e.g. slates, tiles.
- Strict observation of access/egress should be followed during the demolition phase.
- A safe place of work must be provided, e.g. not the top of an unprotected wall.

In the case of steel-framed buildings:

- All non-structural material should normally be removed before any dismantling takes place.
- Main structural members should be supported while the ends are either cut or unbolted. Care is needed as the members might spring away from the structure. The area below should be protected and sealed off if falling material is allowed to drop. On completion, members and/or trusses should be lowered carefully to the ground. The cutting or unbolting operation should be carried out from appropriate work platforms, with operatives wearing appropriate safety harnesses.

Machines

Piecemeal demolition can also take place using machines. Certain precautions are needed:

- Where two structures are attached, they should be separated by a minimum of 1m prior to using a machine.
- Only the machine operator and banksman should be allowed within the restricted area around the demolition site.
- The machine should be adequately protected by being of robust and suitable construction and containing shatterproof material and appropriate strong mesh steel bars/grille.
- The operators must be protected by respirators when asbestos dust or other health hazards may be present.

Machine demolition techniques include:

- **Demolition ball** - 'wrecking ball' - a pear-shaped ball of around one tonne, suspended from a crane, fitted with an anti-spin device, i.e. a dragline. The ball is allowed to fall vertically onto the structure or by swinging it forward or sideways into the structure. Buildings should be less than 30m high before this technique is used.

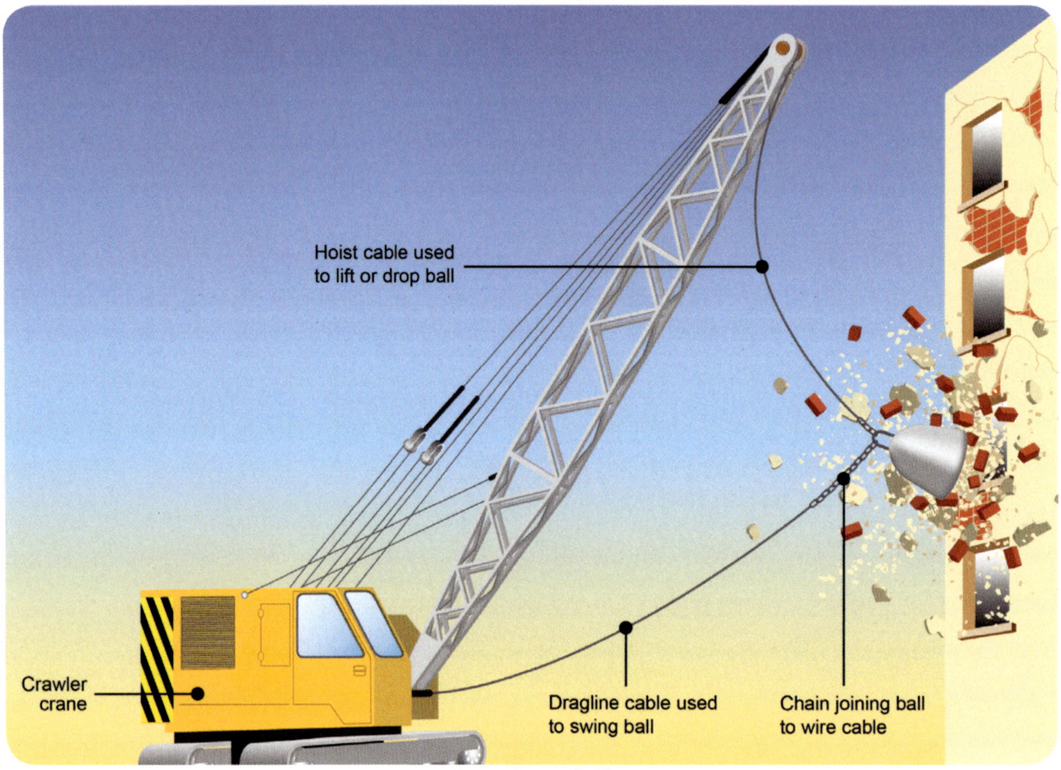

Demolition ball

- **Impact hammers**, which have a heavy duty pick which is incorporated within a track- or wheel-mounted chassis, with an articulated boom. The pick is vibrated using either pneumatic or hydraulic power, and is usually used to remove material from the top of columns or walls (usually limited to 600mm thickness) after removing any steel reinforcement which may be in place.
- A **360° excavator** with multi-functional attachments - these machines have a reach of between 15m and 35m plus the length of the fitted attachment.

A 360° excavator with protective cage

- **Remote control** demolition of a high-level concrete floor - achieved using radio-controlled machinery from a safe distance away.
- Use of a **hydraulic pusher arm**, which is normally mounted on a tracked or wheel chassis and has a toothed plate or hook. The pusher arm pushes into the structure that is being demolished. The equipment does, however, have a limited reach.
- Use of **shears** in cutting steel structures.

Chimneys, pre-stressed concrete, brick and masonry viaducts require more specific precautions to remove the structures in smaller sections, not only to contain the materials during a fall but to retain the fall within a tightly controlled area, such as an exclusion zone (due to the height and length of such structures).

Hazards and Control Measures Relating to Demolition and Dismantling

Demolition is one of the most dangerous activities undertaken in the construction industry. It involves a number of legislative requirements including the **Health and Safety at Work, etc. Act 1974** and the **Environmental Protection Act 1990**. The **CDM Regulations 2015** apply to all demolition and dismantling work, however, a project is only notifiable to the Health and Safety Executive (HSE) if it will last longer than 30 days or involve more than 500 person-days of work.

Special requirements need to be applied where appropriate, e.g. demolition may be subject to the **Control of Major Accident Hazards (COMAH) Regulations 2015** or the **Work at Height Regulations 2005**.

Hazards

Premature Collapse

This is a main cause of accidents in demolition, a common factor being failure to plan the operation at an early stage. Lack of planning can lead to workers devising their own means of access and methods of work, both of which are inherently dangerous. In addition, other works which fall outside the scope of demolition can also threaten the stability of structures.

Buildings and structures can become unsafe during the demolition process and temporary scaffolding/support may be required, e.g. for heavy overhanging cornices. All temporary structures must support the loads required to ensure overloading does not occur. Similarly, structures within buildings may become unsafe, e.g. staircases.

The removal of certain parts of the building or structure during demolition can result in other parts becoming unsafe for persons in, on or near the building being demolished.

Falls and Falling Materials

Falls are often due to poor access and egress, unsuitable pedestrian and traffic routes, and unsafe structures. Work at height is also a major contributor.

Causes of falling materials include:

- Deliberate dropping.
- Materials falling or being ejected from the apertures.
- Inadvertent falls of dislodged material.
- Materials acting as a projectile.
- Falling loads or equipment.

- Lifting loads over areas of public access.
- Unplanned falls of material, including accidental falls of tools and materials.

DEFINITION

DELIBERATE DROPPING

In demolition, deliberate dropping refers to deliberate pulling down or knocking over of a structure (e.g. knocking a wall over with a digger bucket).

Falling materials may also be a risk to operators of vehicles and plant are being used on site.

Plant, Vehicles and Other Equipment Overturning

Transport-related accidents in demolition work are due to:

- Poor, unstable and uneven ground.
- Poor control of vehicle movements.
- Use of unsuitable plant or vehicles.
- Unauthorised use of vehicles and plant by untrained operators.
- The potential proximity of private or road-going trade vehicles to heavy demolition plant.
- Vehicles left in unplanned areas, causing obstructions and obscuring sight lines.

Manual Handling

Damaging work postures are likely in demolition work, involving stooping, kneeling and working in postures with the arms raised above shoulder height. Because of this, the industry has a high incidence of musculoskeletal disorders (MSDs) affecting the lower back, knees, shoulders and neck. In each case, long-term disability is common.

Many of the manual handling risks can be reduced through better planning, control and management. At the practical level, this will involve better co-operation between designers and contractors.

A hierarchy of control measures includes:

- Mechanical means, such as hoists and chutes rather than manual methods.
- Lifting and handling aids such as forklifts, trolleys or barrows rather than carrying.
- Correct choice of job-specific equipment which is well maintained.
- The use of lighter bags and loads.
- The avoidance of repetitive handling and awkward movements.

In Element 7 we will discuss hoists in more detail. Scaffold hoists are good examples of equipment which can be used to avoid excessive manual handling in demolition work.

CASE STUDY

Manual Handling in Demolition Work

A multi-storey office block was being partly demolished and the concrete skeleton was to be incorporated into the new office block. Workers used wheelbarrows to move waste materials and tip them down an open lift shaft. The HSE served a Prohibition Notice on the unguarded open lift shaft edge and an Improvement Notice for the manual handling tasks.

The site management guarded the lift shaft and installed a conveyor system to take materials to the shaft for disposal.

This solution removed the risk of:

- falling down the shaft;
- being struck by falling debris; and
- MSDs;

while increasing productivity.

Dust and Fumes

Dust is a common hazard during demolition. The dust may be an irritant, toxic or corrosive; cause eye injuries; or carry fungal or bacterial matter.

Dusts:

- Are produced mechanically by grinding or similar actions and are divided into two categories dependent on the size (inhalable and respirable - discussed in Element 12).
- Causing damaging health effects are below 10µ in diameter.
- Below 7µ (respirable dusts) are the most damaging to the lungs and are likely to cause chronic respiratory disease.

The dusts generated during active demolition work often contain silica compounds, waste cement, plaster products and possibly Asbestos-Containing Materials (ACMs).

Vehicle movements on dry soil surfaces may also create dusts. Such dust becomes a visible nuisance to local residents. It can prevent people from opening windows or using their gardens. Regular damping of traffic routes can reduce the risk.

Fumes are generated by heating and melting metals from demolition methods that involve welding and burning in the dismantling of steelwork. Techniques generating fumes include the use of plasma arc torches, arc cutters and oxy-acetylene torches.

Noise and Vibration

Demolition is inherently noisy, and uses equipment that can create vibration. As well as causing hearing loss and vibration-related injuries to workers, noise and vibration can be a nuisance to the surrounding community. In such cases, the local authority environmental health department can intervene to reduce it. They should be informed of the progress of the demolition and the dates and times of any particularly disruptive activities and try to address complaints before any formal action is taken.

Existence of Services

Before any demolition work can commence, all services must be identified and the corresponding disconnections built into the demolition plan (see later).

All underground and overhead services (electricity, gas, water, etc.) must be located, as well as services within buildings (intrusion into unpurged pipework for some services may carry risks of explosion):

- Service providers (electricity suppliers, lighting companies, etc.) should be contacted to obtain accurate plans of supplies. Other tell-tale signs, such as junction boxes, should be observed. Cables should be avoided at the planning stage.
- Similarly, water, gas and telecommunications companies should be able to provide plans of their respective buried services.
- Demolition supervisors must be supplied with this information, and should in turn ensure it is passed on to the employee or contractor carrying out the work. Information given by the owner of the site should only be taken as a guide.
- Cable locating devices, in conjunction with cable plans, can subsequently be used to trace the position of cables (but, even where cables carry live current, the detector may still be confused by other close services).
- Suitable pipe-locating devices should also be used, but care should be exercised as these devices will not detect plastic pipes.
- Operators of detection devices must be:
  - Fully conversant with the instruments.
  - Trained in their use.

Hazardous Substances

Hazardous substances can often be found in buildings which are to be demolished or deconstructed. The ground may also carry risks from contamination by hazardous substances. The presence of dangerous substances such as asbestos, polychlorinated biphenyls (PCBs, found mainly in electrical equipment), lead paint, flammable liquids, any unidentified materials, drums, packages, radioactive materials and residual material in pipelines or tanks can pose serious hazards to workers. There can be serious health hazards associated with these, e.g. toxic, carcinogenic, pathogenic (causing disease), sensitising, corrosive or irritant, or radioactive.

Drums of hazardous substances and contaminated ground are often found on demolition sites

The method statement and demolition plan should outline:

- How the presence of hazardous substances is to be identified.
- The means of disposing of such substances.
- The requirements for any protective equipment.

Where existing plant has contained flammable materials, special precautions must be observed to avoid fire and explosion. It may be necessary to engage the services of a competent analyst to identify any residues, carry out monitoring and assess whether any pockets of contamination remain.

Where it is necessary to enter plant for cleaning or to carry out an assessment, breathing apparatus may be required and a permit-to-work system used.

Dilapidation

Some buildings and structures, by the nature of their age and periods left empty and unused, will be in a state of dilapidation (i.e. a state of disrepair or ruin). Dilapidation can add to the hazards of the demolition work by leaving services exposed or hidden by debris and rubbish. Such debris also hides holes and voids, adding danger to the already long list of hazards. The effects of dilapidation will be looked at in the pre-demolition survey (see later).

Controls

While it is the direct duty of the contractor to control the hazards on site, designers should give consideration to measures which remove the hazards altogether or reduce the likelihood of them occurring.

Avoidance of Premature Collapse

DEFINITION

BUTTRESS

An inclined tube assembly anchored at one end (on the ground or other suitable point) and fitted to the scaffold at the other, to widen the structure base and increase stability.

Designers should consider structural stability and include relevant information with their designs in order to avoid premature collapse of structures and maintain structural stability throughout demolition work:

- **Temporary structural supports** should be used in accordance with the appropriate standards and should be suitable for the work required. These include temporary structural support for shoring, propping and facade retention, including standard scaffolding support systems and screw-jack propping (commonly called 'acros'), shoring systems, facade retention systems, and buttresses.

  Temporary structural supports should be:

  - In position before disturbance or removal of the existing supports.
  - Founded on secure and effective footings.
  - Adequately laced and braced.
  - If appropriate, capable of resisting dynamic loadings.
  - Checked both prior to and during demolition for effectiveness, including for the effects of vibration.

  Where scaffolding is required, it should normally be an independent tied scaffold. It must conform to the usual construction standards and, as work progresses, be removed in a progressive and planned manner to ensure its integrity.

  Any shoring and propping should be placed in position at the appropriate time and designed not to interfere with subsequent work. Shoring should be checked for effectiveness as the demolition proceeds and should never be loaded in excess of the design limits.

- **Facade retention systems** should be installed as demolition progresses, to ensure that stability is always maintained. Structural engineering advice should be obtained for the design of temporary facade supports, taking account of the stability of the facade and of the shorings as the demolition proceeds.

  - The method of fixing the facade retention system should not damage the facade.
  - Account should be taken of movement of the facade that could occur before, during or after demolition operations.

- All fixings should be regularly checked throughout the period of the contract.
- Demolition work should be carried out so as not to damage the facade.
- Elements to be removed which are built into the facade structure should be carefully cut away.
- The temporary support system should not be weakened by removal of components until the facade in that location is fully supported either by the new structure or by an alternative temporary structure.

Protection from Falls and Falling Material

Exclusion zones (see earlier) should be applied to all demolition operations, keeping people away from areas of deliberate dropping, the potential for falling materials or inadvertent falls of dislodged material. A safe place of work and safe access in the workplace are to be maintained at all times, including access by roads, gangways, passageways, passenger hoists, staircases and ladders. Scaffolds, permanent access stairs or catwalks (narrow bridges) used by demolition personnel should be maintained in a safe condition, for as long as possible.

TOPIC FOCUS

To **prevent falls**, safe access routes and working places should include:

- Removal of obstructions, including trip hazards such as debris, materials and tools.
- No nails projecting from timbers (remove or hammer flat to remove trip hazard).
- Walkways and working platforms that are not overloaded.
- Pedestrian routes that are segregated from routes used by vehicles and plant.
- Erection of adequate barriers or other edge protection, to prevent falls from the open edges of roofs, buildings, gangways, staircases and lift shafts.
- Holes and openings that are securely fenced off or provided with adequate, fixed, clearly marked and properly secured covers.
- Adequate artificial lighting when work is carried out in the absence of sufficient natural light.

When work at an open edge is the only option (e.g. not using remote machine methods), a safe, independent working platform, or anchored fall restraint or fall arrest equipment, should be used.

Protecting the public

People can be protected from **falling materials** by:

- Making walkways from fully sheeted scaffolding with debris netting or reinforced plastic sheeting.
- Using other temporary structures (e.g. tunnels, protective gantries, fans and crash decks).

Such structures should be designed to take account of the loads (including wind loads) imposed when sheeting and/or netting material are used.

Further protective measures are:

- Fire exit routes must not be obstructed, and possibly be protected with fire-retardant sheeting and netting.
- Where appropriate, windows should be boarded up to prevent materials falling or being ejected from the apertures.

- Personal Protective Equipment (PPE) (hard hats, boots, specialised clothing, goggles, etc.) is to be worn at all times.
- Avoid lifting loads over areas of public access (if unavoidable, temporary closure of the access and the provision of a safe alternative is required).
- Restrict or prohibit access beneath the demolition work at times when safety cannot be guaranteed.
- Remove occupants from a building undergoing major demolition work. If part is to remain occupied, a Safe System of Work (SSOW) is required.

The locations of temporary supports must be identified before demolition starts, and this should be clearly shown in the demolition method statement. No part of the structure should be so overloaded that any part of it becomes unstable during demolition work.

Protective fans, crash decks and gantries are temporary structures that should be of adequate strength to safely prevent premature collapse, including accidental falls of tools and materials. If necessary, they should be waterproofed to protect persons below from drips and should not be used for storage or access.

Where there is a risk of falling materials striking vehicle and plant operators, e.g. excavators used in demolition work, vehicles should be fitted with Falling-Object Protective Structures (FOPS).

Siting and Use of Plant, Vehicles and Other Equipment

Various items of plant and equipment may be used in demolition. The following measures can be taken to avoid creating additional hazards:

- An inventory of the plant and equipment intended to be used on site should be kept, along with any appropriate test certificates.
- Appropriate, up-to-date training certificates for vehicle and plant operators should be kept.
- The plant and machinery should be properly sited within a compound and isolated at the end of each shift.
- Sufficient and suitable parking areas should be provided for all vehicles using the site - including employees' and visitors' private vehicles.
- Vehicles should not be left in unplanned areas, otherwise the safe operation of the site may be jeopardised - for example, by obscuring sight lines.
- Safe areas for loading and unloading skips and collection vehicles should be provided.
- Prevent parking of vehicles near any scaffolding, stacks of materials or compressed gases, or storage compounds containing toxic or flammable substances, especially transport carrying Liquefied Petroleum Gas (LPG).
- Where the risk of vehicles overturning is significant, e.g. dumpers used on sloping ground, vehicles should be fitted with Roll-Over Protective Structures (ROPS).

Dust and Fumes

Protection against dust and fume from demolition activities may include the following:

- A **Control of Substances Hazardous to Health Regulations 2002** (**COSHH**) risk assessment.
- Maintaining a high standard of housekeeping, regularly cleaning and sweeping the site roads.
- Using techniques that produce low levels of dust; damping down.
- Designating vehicle routes to and from the site to limit the problem.
- Maintaining plant and equipment to a high standard, so that it is less likely to break down and will emit fewer pollutants, and to prevent spillage of oil and fuel to the environment.

- Using wheel-washing facilities at exits onto public traffic routes.
- Using water sprays as a dust suppressant, taking care as to the potential consequences of environmental contaminants in the resultant run-off, and the proximity to electrical services.
- Using air movers or LEV where possible.
- Using covered wagons and skips and restricting the speed of vehicles.
- Using PPE as a last line of defence.
- Where asbestos might be present, take samples to verify its presence and carry out the necessary precautions, actions and required notification.
- Put precautions in place if lead is found to be present.

Noise and Vibration

Precautions are required to avoid noise and vibration and provide protection from them for people working in or living near demolition sites:

Workers need to be protected from noise

- A noise assessment may be required, in accordance with the **Control of Noise at Work Regulations 2005**, which will identify all potentially noise-sensitive areas on the site.
- Background noise should be monitored prior to work beginning at any noise-sensitive areas.
- Actions should be taken to reduce noise levels (in the demolition method statement).
- The following should be used:
  - Plant and equipment with low operating noise levels.
  - Mains-generated electricity in preference to diesel generators.
  - Solid-panelled fencing around the site instead of wire fences.
- Deliveries should be restricted to daytime working hours, i.e. not early morning or late at night.
- Project contact details should be prominently displayed in residential areas, to allow any nuisance issues to be addressed quickly.
- Warning notices should be posted to warn people to keep away from the danger area.
- Appropriate silencing/soundproofing should be fitted to all machinery and plant on site.
- Personnel working within the noise zone should be provided with, and use, adequate hearing protection.
- Continuous vibration levels must not exceed 10mm/sec at any nearby buildings (e.g. tower blocks, office blocks, steel-framed factories, warehouse units and structures related to transport, power and water services) and 2.5mm/sec for nearby houses (including houses converted into flats), listed buildings or ancient monuments.

Protection of the Environment

Contaminants from construction work must be closely controlled to prevent damaging releases into the environment (onto land, or into air or water) during both demolition and transportation for disposal.

Environmental information relating to any underground tanks or potential chemical/biological hazards or the presence of contaminated land, along with the building's former use, should be supplied to the contractor prior to demolition. Environmentally damaging releases can occur from:

- Waste disposal sites, gasworks, power stations, sewage treatment works.
- Railway land including sidings and depots, coal-yards (contaminated by phenols).
- Oil refineries, petrol stations and associated sites, metal mines, smelters, electroplating works, refineries and foundries.
- Chemical works.
- Industrial Revolution era workshops and warehouses.
- Slaughter yards and abattoirs, fish and meat markets, tanneries.
- Paper and printing works, wood preservative treatment sites.
- Hospitals, including isolation and quarantine premises.
- Buildings with animal hair used as insulation in roofs, etc. (anthrax).

Protection is also required from noise, dust and site waste, and suitable controls should be identified in the risk assessment and included in the health and safety plan for the demolition project.

STUDY QUESTIONS

1. In addition to piecemeal demolition and deliberately controlled collapse, identify three other methods of demolition.
2. Identify three common causes of transport-related accidents in demolition work.
3. With regard to hazardous substances, what three significant areas should be outlined in the method statement and demolition plan?
4. How is the premature collapse of a building/structure avoided during demolition?
5. What protection measures are used for noise during demolition?
6. What protection measures are used for dust control on demolition sites?
7. What precautions are put in place to ensure the safe siting of plant and machinery on a demolition site?
8. How is the environment protected during the demolition process?

(Suggested Answers are at the end.)

Purpose and Scope of Pre-Demolition, Deconstruction or Refurbishment Survey

IN THIS SECTION...

- Owners of property and controllers of premises have duties for carrying out pre-demolition surveys.
- Only competent persons can conduct pre-demolition surveys.
- Structural elements and location of services must be identified, as well as effects of dilapidation on the structure.
- Drawings, calculations and the health and safety file should be reviewed as part of the survey.
- All previous alterations to the structure are to be taken into account.

Pre-Demolition Investigation and Survey for both Notifiable and Non-Notifiable Demolition Projects

The purpose of a pre-demolition investigation and survey is to collect all relevant safety, health and environmental information relevant to the site, to establish an SSOW and a demolition plan.

A pre-demolition survey

The survey is performed by an initial desktop study followed by an on-site survey, taking into account any off-site features that can affect work on site.

The investigation and survey should consider a number of issues:

- The **type of structure and method of construction** of the building and **the condition** of structural members, roofs, walls, floors; variations in building construction where old and newer buildings may be present; and the general stability features of the structures and buildings.
- The **presence of cellars**, basements, vaults, underground tanks, confined spaces, etc. Many of these features can remain hidden by the debris of dilapidation and age.
- The **identification of services**, above and below ground (electricity, gas, water). Again, dilapidation can hide the presence of services, or expose them in critical locations.
- The **presence of hazardous substances** or contaminated land (see earlier).

  As discussed previously, where asbestos might be present, samples need to be taken to verify its presence and the necessary precautions, actions and required notification should be carried out; and precautions need to be put in place if lead is found to be present.
- **Waste management** should ensure that wastes are separated on site into their different waste streams and controlled separately. Licensed waste removal contractors will have to dispose of the waste at licensed sites. Some of the waste, such as asbestos, will be controlled waste which is required to be kept in locked skips until disposal off site.
- **Safe access** onto and around the site must be maintained at all times, together with safe egress, for vehicles and pedestrian traffic. Access must always be maintained for emergency vehicles.
- The **proximity and condition of other structures**, e.g. hospitals, schools, care homes, industrial premises, roadways, waterways, bridges, etc.

Other useful information that would complement the demolition plan could include:

- The site's physical features and hydrology.
- The systems (shoring or other supports) necessary during demolition.
- The method of bonding of the main load-bearing walls.
- The possibility of making the adjacent buildings watertight.
- The effects of removing superstructure (an upward extension of an existing structure), stabilising loads, additional props or support during the demolition.
- The actual sequence of demolition operations (reverse of erection of the structure if possible, i.e. the preferred demolition work method is by reducing the height of the building initially and then carrying out a controlled collapse so that work can be completed at ground level).
- Site storage facilities; compounds for machinery, fuels, oils, etc.
- Waste disposal, removal and storage of rubble from the immediate area during demolition and its eventual removal off site (possible reuse of rubble for building up towpaths, etc.).
- The proximity of people to the site, e.g. access/egress for vehicles, noise, vibration, dust problems, protection of the public.
- Timescale of the demolition job to minimise interference for other jobs on site and the public.
- Establishment of site safety following demolition, e.g. safety of the remaining structure, control and disposal of debris, preventing unauthorised access, safety relating to gravity chutes used.

Duties of the Property Owner

The **property owner** or person in control of the premises has a duty to ensure that a pre-demolition survey is carried out on non-domestic premises, whether the project is CDM notifiable or non-CDM notifiable. In the case of refurbishment or deconstruction, the local authority building department should be consulted before any structural alterations are made.

A refurbishment or demolition survey is necessary when the building, or part of the building, is to be upgraded, refurbished or demolished.

Pre-demolition, deconstruction or refurbishment surveys should be carried out by **competent persons**, usually being structural engineers, asbestos surveyors, etc.

The expertise of a structural engineer must be used in the following cases:

- In the design of a facade retention scheme.
- Where there is doubt:
  - Over the building's stability.
  - About the proposed method of demolition.
  - About the capacity of the building to take loadings.

It is good practice to consult a structural engineer at the planning stage of demolition to avoid uncontrolled collapse.

The **purpose of the survey** is to carry out a thorough structural investigation and assessment before any potentially load-bearing parts of a structure are altered. It will also determine the presence (or absence) of asbestos (what types and in what locations). The details of the survey are covered in the next section.

Identification of Key Structural Elements Including Pre- and Post-Tensioned Components

Structural collapse is to be prevented, so all key structural elements of the building are to be identified, such as pre- and post-tensioned components.

Pre-tensioning is where steel cables are placed under tension (stretched) and the concrete is poured around them until everything sets. When the concrete has hardened, the machine used to tension the steels is released. The tension in the cables now places the concrete under a compressive force increasing its strength. Post-tensioning is where steel cables are set in sleeves and concrete poured over them. Tension is then applied to the cables after the concrete has hardened. Examples of structures where pre- or post-tensioning techniques have been used are bridges and multistorey car parks.

The competent person will decide on the suitable method of demolition and the design and nature of any temporary structural supports that may be required - they are to withstand foreseeable loads. During the work, structures are never to be overloaded.

Identification of Location and Type of Services

The location and identification of any services are to be determined, such as electricity, water, gas, etc. Either integral, underground or overhead services are likely to be encountered in demolition, refurbishment or deconstruction projects.

Identification, Significance and Extent of Any Dilapidation of the Structure

The extent of any dilapidation is to be determined, identifying the locations and significance of its effects on the structures involved in the work.

Review of Existing Documents

A review should be made of:

- any existing drawings including those available from service providers;
- structural calculations such as load bearing capacity of floors; and
- the existing health and safety file for the project;

together with further information that is available and that may be relevant to the work in hand.

This will give an indication of the effects that the demolition, deconstruction or refurbishment may have on the structures and nearby structures.

Review of All Structural Alterations Carried Out on the Structure in the Past

All previous alterations to the structure should also be reviewed and compared with the health and safety file, calculations, drawings, etc., to ensure the significance of any change does not adversely affect the current project work.

STUDY QUESTION

9. What information is to be considered in a pre-demolition survey?

(Suggested Answer is at the end.)

Summary

This element has dealt with safety in demolition activities.

In particular, this element has:

- Introduced basic terminology and methods used in demolition and dismantling.
- Identified the main hazards associated with demolition and dismantling work: premature collapse; falls and falling materials; plant, vehicles and other equipment overturning; manual handling; dust and fumes; noise and vibration; existence of services (buried, overhead and in buildings); hazardous substances; and dilapidation.
- Outlined the control measures for demolition and deconstruction: avoiding premature collapse; protection from falls and falling material; siting and use of plant; how to control dusts, fumes, noise and vibration; and protection of the environment.
- Identified the purpose and scope of pre-demolition/refurbishment surveys, including: the duties of owners of properties and competent persons to carry out investigations; identification of key structural elements, identification of services and the significance of dilapidation of structures; the need to review drawings, calculations and the health and safety file, and all structural alterations carried out prior to the current project.

Exam Skills

Question

Scenario

An old manufacturing facility is to be demolished in order to make way for a new development. The site is on an existing industrial area on the outskirts of a town. The facility is close to a busy highway with a river at the back of the site.

Task: Demolition Survey

You have been asked to conduct a pre-demolition survey of the facility. Prior to starting the survey, you decide to make notes on key issues to be addressed to act as an aid during the demolition survey.

Prepare the notes you will use for the demolition survey.

(Total: 10 marks)

Approaching the Question

Now think about the steps you would take to answer this question:

Step 1 The first step is to **read the scenario carefully**. Note you are about to conduct a demolition survey so think about relevant things that would need to be covered. Even though they have not been mentioned in either the task or the scenario, your analytical approach guides you to use examples.

Step 2 Now look at the **task** - prepare notes (and examples) on:

- Significant hazards such as premature collapse.
- Services that might exist and require isolation, e.g. electricity/gas.
- Selection of appropriate demolition methods.

Step 3 Next, consider the **marks** available. In this task, there are 10 marks so it is expected that around 5 or 6 different pieces of information should be provided. The task does not split the available marks – you will need to consider how the marks will be applied, so probably 2 marks for each piece of information.

Step 4 **Read the scenario and task again** to make sure you understand them and have a clear understanding of the demolition process. (Re-read your study text if you need to.)

Step 5 The next stage is to **develop a plan** - there are various ways to do this. Remind yourself, first of all, that you need to be thinking about 'considerations in a demolition plan and not the hazards associated with demolition'. Remember you are creating notes, not a bullet-pointed list. However, the answer plan can be a bullet-pointed list that you use to develop into notes for your meeting.

Suggested Answer Outline

Demolition survey key elements:

- Key structural elements.
- Services.
- Dilapidation.
- Review.
- Structural alterations.
- Selection of appropriate method.

Now have a go at the question yourself.

Example of How the Question Could be Answered

The purpose of a demolition survey is to ensure hazards associated with the general structure (load bearing parts), the presence of materials (asbestos) and location of services (electricity and gas). Considerations of the survey need to take into account the following:

1. *Identifying key structural elements in order to prevent unplanned collapse. A suitable method of demolition and the design and nature of any temporary structural supports can then be decided.*
2. *The location and identification of any services such as electricity, water and gas. The services might be integral, underground or overhead.*
3. *Identifying the extent of dilapidation is important as this might affect the structure involved in the work.*
4. *A review should be made of the health and safety file for the project. This should take into account structural calculations that may be relevant to the work in hand. This will give an indication of the effects the demolition may have on the structure and nearby structures.*
5. *Comparison of previous alterations to the structure need to be reviewed and compared with the health and safety file. This is to ensure any changes made to the structure do not adversely affect the current project work.*
6. *The final part of the demolition survey is to design a method of demolition. This will be based on the type of structure and the information gained from the survey.*

Reasons for Poor Marks Achieved by Exam Candidates

- Not expanding the answer beyond a few words as opposed to giving a sentence of explanation.
- Misinterpreting the question by identifying demolition hazards and control measures associated with demolition activities, as opposed to considering the content of a demolition survey.
- Focussing on two or three areas and going into too much detail opposed to going wider covering all relevant points.

Element 6

Mobile Plant and Vehicles

Learning Objectives

Once you've studied this element, you should be able to:

1. Explain the hazards and control measures for the safe movement of vehicles/ plant within a construction environment, including when using public highways as a workplace.
2. Outline the factors associated with driving at work that increase the risk of an incident and the control measures to reduce work-related driving risks.

Contents

Safe Movement of People

IN THIS SECTION...

- When considering the movement of people on and around a construction site, we must take into account construction hazards including being struck by falling or moving objects, collisions with moving vehicles and striking against fixed objects.

Hazards to Pedestrians

Striking by Moving, Flying or Falling Objects

Such collisions may be the result of accidents themselves, or the consequences of unprotected moving parts being too close to pedestrian walkways/pathways.

- Typical **moving object** hazards include:
  - Normally secured objects breaking loose and hitting people, e.g. a container sliding down a slope.
  - Material being transported on site.
  - Loads being moved on cranes.
  - Automated machinery.
- **Flying object** hazards:
  - Are often the result of an accident, e.g. parts being ejected from a machine or a broken drill bit coming off.
  - May be from normal work activities, such as waste materials flying out from drilling or cutting operations, e.g. when stone, concrete or rock are being broken up.
  - Roofing sheets and plywood could fall and become flying objects if not secured properly, especially in high winds and if stored at height.
  - Thrown objects, such as a scaffold clamp.
- Being hit by **falling objects** is a common hazard, particularly on construction sites. Any unsecured object near an edge or on a sloping surface may be liable to fall and injure someone passing or working below, for example:
  - Materials on shelves or the tops of cabinets.
  - Pallets and badly stored bricks.
  - Unsupported roof trusses propped against a wall.
  - Loose materials on scaffolding or working platforms.
  - Deteriorating brickwork or where edges are unprotected.

  Falling objects could be caused by a number of things, including:
  - Unstable access equipment (which can collapse or overturn).
  - Poor housekeeping and incorrect methods of getting materials, equipment or tools from ground level to the working area or vice-versa.
  - Adverse weather conditions, e.g. high winds.

There are many opportunities for accidents involving people and vehicles on a construction site

- Workers accidentally dropping things.
- Poorly secured or stacked objects.
- Wear and tear.
- Unprotected edges on scaffolding.

Collisions with Moving Vehicles

Any situation which involves crossing a vehicle traffic route is a hazard, particularly where visibility is restricted and it is not possible to see (or hear) the approach of vehicles clearly.

Emerging from doors can present a problem where the exit leads straight onto a vehicle path, or if vehicles are obscured by building debris, stacks of scaffolding, etc.

Other hazardous situations include:

- Working in proximity to moving vehicles, e.g. in loading bays or carrying out road repairs.
- Where vehicles cross or move around areas which are normally reserved for pedestrians.
- Where reversing or refuelling is necessary.

Striking Against Fixed or Stationary Objects

This may occur as a result of objects projecting out into areas where people walk or simply being placed in a position where people may walk or run into them, such as:

- Poor stacking may leave pipes, rods, projecting reinforcement bars or pieces of timber sticking out into pedestrian walkways.
- The edges of scaffolding, or even the cross-poles themselves, may be at a level where people might walk into them.
- Bricks, boxes and containers of materials may be left outside buildings, etc. awaiting removal or use.

Any of these situations may be made more dangerous where visibility or movement for the pedestrian is restricted.

STUDY QUESTION

1. State four reasons why objects may fall on pedestrians.

(Suggested Answer is at the end.)

Safe Use of Vehicles and Plant

IN THIS SECTION...

- Vehicles and plant present a number of hazards to all persons on a construction site. These hazards include:
  - Hazards causing loss of control and overturning of vehicles.
  - Collisions with other vehicles, plant, pedestrians and fixed objects.
  - Non-movement-related hazards, e.g. loading, unloading and securing loads, sheeting and coupling.
- Vehicle operations on a construction site must be controlled to reduce the risks to persons working or visiting. These controls include:
  - Safe site: suitability of traffic routes, management of vehicle movement (including using banksmen) and environmental aspects such as visibility, gradients, changes of level and surface conditions. Also keeping vehicles and people apart, protecting them with barriers, signs, warnings, etc.
  - Safe vehicles: vehicle selection, and vehicle inspection and maintenance.
  - Safe drivers: selection and training of drivers and management systems for ensuring driver competence.

Hazards from Workplace Transport Operations and Plant

DEFINITION

MOBILE WORK EQUIPMENT

- Includes dumpers, forklift trucks, mobile cranes, land rovers, concrete wagons, etc.
- Does not include pallet trucks, sack barrows, wheelbarrows, compressors and concrete pumps.

The movement of vehicles represents one of the biggest hazards in the workplace - in respect of both the risk of injury to people and of damage to buildings, plant and equipment. Of the many fatal accidents in the construction industry, those involving moving vehicles, including mobile work equipment, are among the most common.

If not well controlled, vehicle operation can lead to loss or injury resulting from loss of control, overturning or collisions with other vehicles, pedestrians, plant or fixed objects. In addition, there are non-movement-related hazards to consider such as loading, unloading, securing and sheeting loads; coupling; and vehicle maintenance work.

To identify the hazards presented by vehicle operations, you need to take account of the:

- Required patterns and volume of vehicle movement in and around the workplace, including access to and egress from it to the public roads.
- Types of vehicles on the premises.
- Types of operations undertaken by vehicles in the workplace and the types of load being moved.
- Need to segregate vehicles and pedestrians.
- Impact of different weather conditions.
- Local environment - condition of the roads, lighting and sight lines.

Movement-Related Hazards

Loss of Control

Assuming driver competence, loss of control is likely to be caused by:

- **Mechanical failure**, e.g. of the braking or steering mechanisms, causing the driver to be unable to apply key controls.
- **Environmental conditions**, affecting:
  - The road, e.g. excess water, mud, snow and ice.
  - Visibility, e.g. sunlight suddenly blinding a driver.

If a vehicle is out of control it may:

- Hit something.
- Overturn.

Overturning

There are two main ways in which a vehicle may overturn:

- Onto its side - **lateral instability** - as when a high-sided lorry is blown over in high wind, or when a loaded forklift tips over while driving across a slope. It is caused by the centre of gravity of the vehicle moving outside of its wheelbase.
- Onto its front or back - **longitudinal instability** - as when a tractor's front wheels lift due to the weight applied by an attached trailer, or when a forklift truck is moving up or down a slope.

Forklifts with a low and high load causing overturning from lateral instability

Instability is a major problem in the safe use of **high vehicles**, particularly those with **lifting mechanisms**. The risk of overturning is increased by the:

- Speed of travel.
- Steepness of the slope - travelling across a slope.
- Height of the vehicle, including its load.
- Stability of the load.
- Presence of high tyre pressure.
- Presence of external longitudinal pressure, e.g. wind or colliding objects.
- Presence and size of any bumps or holes in the surface.

TOPIC FOCUS

Overturning Vehicles

Reasons why **site vehicles** overturn include the:

- Nature of the ground over which it operates.
- Obstacles on the traffic route/operating area.
- Suitability of the vehicle for the operations.
- Condition of the vehicle/way it is maintained.
- Pressure in the tyres.
- Way the vehicle was loaded (uneven or overloaded).
- Competence of the driver.
- Speed at which the vehicle is operated.

Collisions with Other Vehicles or Plant, Pedestrians or Fixed Objects

These are generally caused by:

- Poor weather - ice or rain causing skidding or longer braking distances, rain or fog restricting visibility.
- Bright sunlight.
- Poor local lighting - decreasing visibility or obscuring parts of the traffic route, including obstructions.
- Inappropriate braking (skidding or longer braking distances).
- Poor sight lines - causing blind spots and concealed junctions.
- Poor ground surface conditions - uneven, muddy or slippery surfaces.
- Congestion - there simply being too many vehicles (or pedestrians) in one area.
- Lack of warning signs and barriers at crossings.
- The roadway being too narrow.
- Inappropriate use of speed.
- Gradients and changes of level.

Non-Movement-Related Hazards

Loading and Unloading

Loads carried by a vehicle often become unstable and either fall off, causing the vehicle to overturn, or cause the driver to lose control. This is due to some deficiency in the loading (or unloading) procedure, such as:

- Overloading.
- Uneven distribution.
- Incorrect positioning or unsecured on forks.
- Unsecured loads.
- Slippery surfaces on the vehicle or load itself.

Securing and Sheeting Loads

Loads must be securely attached to vehicles and sometimes are required to be covered to prevent egress of dirt and dust (known as 'sheeting'). Drivers and vehicle operatives can be at risk of falls from height, falling loads and manual handling while sheeting, netting and securing loads. This could be due to:

- Loads being unstable and shifting while being tied or netted.
- Poor lighting conditions in loading areas.
- Bad weather conditions during operations.
- Inappropriate netting or sheeting access.

Front-to-rear sheeting system

Coupling Vehicles and Trailers/Attachments

Hazards occur if persons are in contact with, or in close proximity to, vehicles and vehicle coupling mechanisms while trailers are being attached to traction units. There is always a possibility of becoming trapped or crushed between a reversing vehicle and the trailer.

Control Measures to Manage Workplace Transport

Safe Site

Suitability of Traffic Routes

All vehicle routes should be clearly marked and separate pedestrians from moving vehicles. The size and layout of routes will be determined by traffic volume.

Traffic routes should allow adequate space for, and be arranged to enable, all forms of movement by the types of vehicle using them, including passing, turning, reversing, loading and unloading.

They must be wide enough and high enough for the safe movement of the largest vehicle likely to use them and make allowance for visiting vehicles which may be larger than those used exclusively on site.

Pedestrian-Only Zones

A pedestrian zone is simply an area where vehicles are restricted and reserved for pedestrians who are free to occupy the entire space. The zone entrances and exits must be clearly designated with signs to ensure all drivers are aware when they are entering or exiting such an area.

Crossing Points

Designated pedestrian crossing points should be clearly marked and controlled by traffic lights if necessary. Where traffic is particularly heavy, bridges or subways may be necessary.

Crossing points must be positioned at places where people WILL cross, not where designers think they should cross. Otherwise, a means of shepherding such as barriers must be erected.

Where walkways cross roadways, a clearly signed and lit crossing point where drivers and pedestrians can see each other clearly should be provided.

Site Layout

The layout of construction sites should be well planned and organised with appropriate areas for safe vehicle operations, including:

- Reversing.
- Manoeuvring.
- Parking.
- Loading and unloading.
- Refuelling.
- Access for emergency services.

Consideration must be given to:

- Locating fire-fighting equipment.
- Storage of plant and materials.
- Workspace for construction activities.
- Site offices and welfare facilities.

Spillage Control

Oil and diesel spills, hazardous chemical spills, fluids including paint and solvent spillages on construction sites can have far reaching implications for people, animals and the environment.

They create a risk of vehicle collision, slips and trips, toxic burns and scalds. They pose a risk to employees, contractors or visitors.

Uncontrolled, spills can get into waterways and soak into groundwater, potentially contaminating animal and human drinking water, threatening plant and water life, wildlife and human health.

A thorough risk assessment will identify the chances of a significant spill and the need for a spillage procedure which will be specific to the substance in question.

The traffic route should be designed with cambers to assist with drainage and appropriate spill kits should be provided along the route.

Management of Vehicle Movements

Controls for vehicle movement will vary from site to site, particularly if there are special considerations which should be taken into account. However, the following controls should be in place on all sites (if applicable).

Reversing

A control hierarchy for reversing could be followed:

- Eliminate/reduce reversing by use of one-way systems and turning areas.
- Ensure adequate visibility (good layout, good lighting, high-visibility clothing).
- Use safe systems of work, a banksman for reversing and difficult manoeuvring, and mirrors.
- Use warning devices (audible and visible).
- Use traffic/warning lights at heavily used junctions, at blind exits, bends or designated pedestrian crossings.

Vehicle Parking

Sufficient and suitable parking areas should be provided for all vehicles using the site - including employees' and visitors' private vehicles.

Signs and Markings

Surface markings and road signs should be used to control vehicle movements.

Signalling - Use of Banksman

Where vehicles are required to reverse, or operate in confined spaces such as loading bays, the use of a person (banksman) to direct vehicle movements may be required. There are standard hand and directional signals used for this, including those shown in the following diagrams.

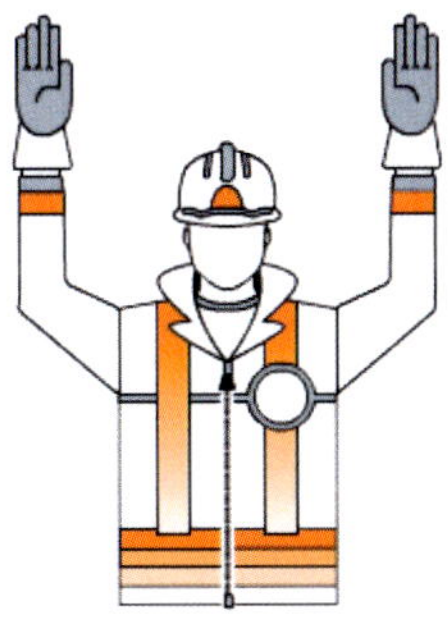

Danger (Emergency Stop)

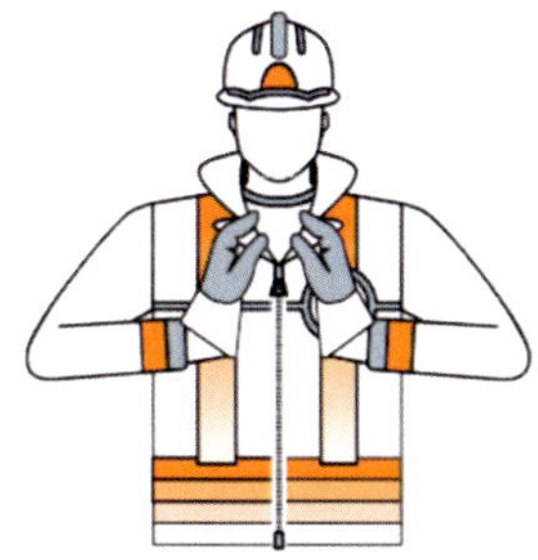

Indicating Horizontal Distance

Move Forwards

Move Backwards

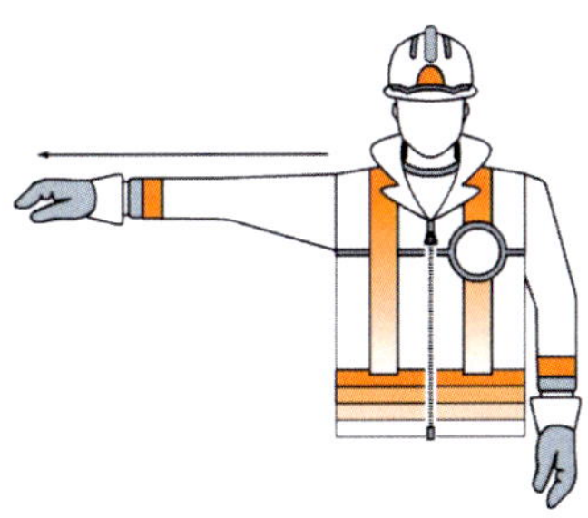

Move to the (Signalman's) Left

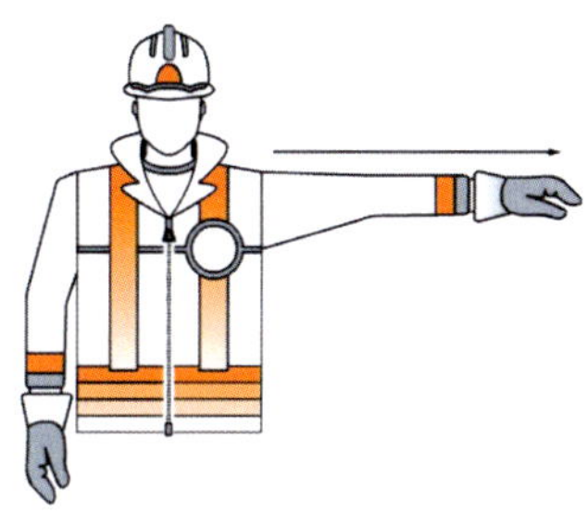

Move to the (Signalman's) Right

Standard hand and directional signals

Loading and Unloading

Many accidents occur during loading and unloading activities. These can be reduced by well-designed work areas and safe systems of work.

TOPIC FOCUS

Loading procedures should ensure that:

- Vehicles are not loaded beyond rated capacities or the legal limits of gross weight.
- The floor of the vehicle is checked for soundness before loading.
- Loads are evenly distributed, and secured or arranged so that they do not move in transit.
- Tail and side boards are closed.
- Overhangs are kept to a minimum and suitably marked.

The following precautions should be taken when **unloading**:

- The load should be checked for stability before ropes or tarpaulins are removed.
- The load should remain stable and evenly distributed as far as practicable during unloading.

Environmental Considerations

Visibility/Lighting

Adequate lighting is essential for safe vehicle operations and security purposes. Outside, areas with natural daylight will require artificial lighting when it is dark and during weather conditions which restrict visibility. Increased levels of lighting may be necessary in certain circumstances, for example:

- At junctions of traffic routes.
- Where traffic routes are close to buildings or plant.
- In pedestrian areas.
- In places where there is regular movement of vehicles and other mobile plant.

To ensure visibility (line of sight) is not impeded:

- Lighting columns should be set back from the road edge. Alternatively, overhead or wall-mounted lights should be used.
- Hazards such as sharp bends or blind corners should be eliminated.
- Walls, stacks, etc. at corners and junctions should be lowered/removed.
- Parked vehicles should not be allowed to cause obstruction to sight lines.
- Site vehicles should use flashing beacons when entering and when in use on site.

Where hazards cannot be eliminated, placing warning signs, hazard markings and mirrors can reduce the risk.

Gradients and Changes of Level

Travelling across gradients or changes of level can be very hazardous, particularly in vehicles designed for smooth routes. To control these hazards:

- Excessive gradients - those in excess of 1 in 10 - should be avoided.
- Steep ramps should always be clearly marked.
- Strict rules about driving on steep slopes are required.

- Vehicles suitable for the terrain should be selected.
- Suitable risk assessments will determine what vehicles can be used safely, i.e. will be stable on certain gradients/ slopes, and highlight their limitations.
- Manufacturers' information about load and stability under operating conditions should be adhered to.
- Normal safe driving practices and procedures should be followed.
- The camber of traffic routes should allow for adequate drainage, but not be any steeper due to the risk of overturning.
- Vehicles that are at risk from overturning should be fitted with suitable Roll-Over Protection Structures (ROPS) - see later in this element.
- Wherever possible, traffic should not be allowed to travel across slopes.

Surface Conditions

The condition of the surface on traffic routes on construction sites is also important in managing risk, in particular:

- Suitable road conditions/surfaces must be provided for all vehicles - delivery lorries, vans, cars and buses.
- All routes should be even, constructed of suitable materials for the vehicles using them, and well-drained.
- The type of surface must be suitably firm, hard and, if damaged, repaired to a high standard.
- Raised kerbs should be kept to a minimum where collision with them may cause overturning, e.g. where lift trucks operate.
- In areas where ground conditions may be unsuitable, improved drainage, compacted hardcore or spreader plates may be used to improve the conditions.

Suitable road conditions should be provided for all vehicles

Iron work such as access covers can be very slippery when wet. **Non-slip coatings** of heavy duty paint blended with fine grip are a quick and easy solution.

Maintenance and Checking of Traffic Routes

Roads are to be well maintained:

- Potholes should not be allowed to develop.
- Loose materials should be cleared regularly.
- Snow clearing and gritting may be necessary.
- Temporary obstructions (including goods, etc. fallen from vehicles) should be cleared as soon as possible.
- Road-sweeping machines should be used to clean the surfaces and maintain them in good condition.

Means of Segregating Pedestrians and Vehicles

Wherever possible, pedestrians should be physically segregated from vehicle traffic routes. Measures to ensure this include:

- Vehicle barriers or the use of clear surface markings to show the separate routes. These are particularly required outside exits from buildings where there is a risk of pedestrians walking directly onto a road.

- Separate access for pedestrians such as a raised or railed-off footpath should be considered where vehicles pass through doorways, under narrow bridges or through narrow tunnels.

In situations where pedestrians and vehicles cannot be separated for operational reasons, the following measures should be taken:

- Pedestrians may be required to wear high-visibility jackets.
- Vehicles should be fitted with warning lights and alarms to indicate movement, particularly for reversing.
- Structural and environmental conditions should be altered to enhance safety, with particular attention to visibility (through lighting and mirrors).

Protective Measures for People and Structures

The following protective measure should be considered:

- Vulnerable plant and materials should be located away from roads or, where this is not possible, suitably constructed barriers should be provided for protection.
- Columns, pillars and walls may need to be padded to limit damage in the case of vehicles colliding with them.
- Barriers may also be used in traffic routes to:
  - Prevent pedestrians entering.
  - Separate vehicles.

Site Rules and Signage

It is important that all drivers using a site are aware of all the rules that apply - one-way systems, speed limits, parking, etc. Visiting drivers should be made aware of these on entering the site.

Safe Vehicles

Rules should be in place to prevent unauthorised use or misuse of vehicles, and to ensure that unattended or parked vehicles do not create hazards, for example:

- Remove keys from vehicles not in use.
- After work, switch off the engine and apply the brake. On battery-operated vehicles, the battery should be disconnected.
- Park in a safe place and do not obstruct emergency exits, other vehicle routes, fire-fighting equipment or electricity control panels.
- Do not leave vehicles unattended on a gradient. If a vehicle has to be left or parked in an emergency, even for a short period of time, it should be left in neutral with the parking brake applied and wheels chocked to prevent unexpected movement.
- Sound the horn at every potential danger point, such as before entering doorways and at blind corners, although the use of a horn does not give the driver right of way.

Vehicle Selection

Checks to make when selecting a vehicle are:

- The environment in which they operate.
- The ground conditions.
- Its stability, e.g. on muddy/soft ground.

- Safety features, e.g. falling/rolling over.
- Lights/warning devices.
- Adequate driver visibility, especially for reversing.
- Driver protection from adverse weather conditions.
- Adequate guarding on dangerous parts.

Vehicle Inspection and Maintenance

Vehicle inspection and maintenance includes:

- Daily checks by the driver.
- Weekly checks (in more detail) by a maintenance engineer.
- Periodic servicing.
- Formal inspections.
- Braking systems.
- Steering mechanisms.
- Tyres.
- Exhaust systems.
- Audible warning systems.
- Driving lights, including flashing warning lights.
- Rear view mirrors and/or convex mirrors to overcome blind spots.
- Roll-over and falling object protection.
- Weather-proof cabins.
- Seat belts and restraining systems.
- Fire-fighting equipment.

Other Considerations

Speed Limits

Often sites set the same speed limit across a whole work area. This is may not be appropriate, and could be difficult to enforce. Usually, different speed limits will have to be set for different routes, because the types of traffic and tasks are different.

The limits that are set need to be appropriate for:

- the vehicles using the route;
- the types of load they carry and how they carry them;
- the driving surface;
- the layout of the route, including how tight the bends are and visibility at junctions;
- hazards along the way; and
- work that takes place on or near the route.

Speed limits need to be clearly signed, especially at points where they change and they must be enforced.

Forklift Trucks

Special rules for these include:

- Lower the forks whenever the vehicle is parked.
- Avoid violent braking of a loaded truck, as sharp movements could cause the load to fall or the truck to tip.
- Whenever possible when driving, lower the forks to within 150mm of the ground and tilt the mast back. (Driving with the load elevated increases the risk of overturning.)
- When a high load restricts vision, drive the truck in reverse, except when driving up a gradient.
- To minimise the risk of overturning, when a loaded truck is travelling up or down a gradient, the forks should face uphill. When unloaded and travelling down gradients, the forks should face downhill.

Vehicle Parking

Rules should be in place regarding on-site parking:

- Parking areas for site and staff vehicles should be separate and have adequate access and egress.
- Space should be allocated for loading and unloading skips and collection vehicles.
- Prohibit parking near any scaffolding, especially vehicles carrying Liquefied Petroleum Gas (LPG).

Safe Drivers

Selection and Training of Drivers

Selection

Drivers must be 17 years old (21 for large heavy goods vehicles) and have passed the necessary test(s).

Selection tests can avoid wasteful attempts to train unsuitable people as drivers. Potential operators should:

- Be reliable.
- Have a mature attitude.
- Be able to do the job in a responsible manner.
- Have a reasonable level of physical and mental fitness.

These parameters do not automatically exclude disabled people, but in such cases it is advisable to seek medical advice to assess their suitability.

Training

This should include:

- **General basic training** - the basic skills and knowledge required for safe operation of the type of vehicle and any attachments which the driver will be required to use.
- **Specific job training** - covering knowledge of the workplace, any special requirements of the work to be undertaken and the use of attachments in specific situations. This will include:
  - Detailed controls of the vehicle to be used.
  - Routine inspections which should be carried out by the operator.

Refresher training is required to operate different vehicles

  - Use of the vehicle in various locations and different weather conditions.
  - Site rules.
  - Loading and unloading procedures and transporting loads.
- **Refresher training** is also beneficial for all drivers, but is specifically required when drivers need to operate different vehicles or if there is a change of work.

The necessary training records must be kept and maintained for each driver containing all the necessary information, e.g. initial/refresher training, any specialist certificates.

Types of Reversing Assistant

Reversing is the most dangerous manoeuvre made when driving. Traditional reversing aids like mirrors are still used on most vehicles and they are very effective if positioned correctly and kept clean. Another aid that has been around since the 1800s is the Fresnel lens - a flattened convex lens that affords a wide-angle downward view of the vehicle.

Relatively modern developments have led to the widespread use of rear-facing cameras and proximity sensors which are now fitted as standard on most domestic and construction vehicles.

STUDY QUESTIONS

2. What are the three main types of hazard caused by vehicle operations?
3. What may cause loss of control of a vehicle?
4. What factors may contribute to the risk of overturning?
5. Identify local environmental conditions that may contribute to vehicle accidents.
6. How can visibility be improved?
7. Identify the main features used to manage vehicle operations and movements.
8. How can traffic routes be maintained to ensure the safety of vehicles?
9. What are the three main types of training available to drivers?

(Suggested Answers are at the end.)

Work-Related Driving

IN THIS SECTION...

- Risk assessment should consider the driver, the vehicle and the journey to be undertaken.
- Road safety must be managed by employers by having a suitable policy, suitable safety management systems, monitoring performance, having a management structure and organisation to manage road safety, and making all employees aware of their legal responsibilities.
- Control measures should be in place to reduce work-related driving risks.

Managing Work-Related Driving

The Health and Safety Executive (HSE) and the Department for Transport have jointly published *Driving at work. Managing work-related road safety* (ING382). It estimates that up to 67% of all road traffic accidents involve someone who was at work at the time.

The approximate cost of an investigation into a fatal road traffic accident (which can be recharged to the company) is £1.4 million. The company image will be tarnished through bad publicity and there may be an increase in insurance costs.

Under the terms of the *Road death investigation manual*, the police establish immediate liaison with the HSE and an investigation is instigated by both. The police have a duty to investigate corporate liability and a parallel investigation is carried out by the HSE; although approval from the Director of Public Prosecutions (DPP) is required for the police force to investigate using the **Health and Safety at Work, etc. Act 1974** (**HSWA**).

In most cases, the police will continue to take the lead on investigating road traffic incidents on public roads and the HSE will usually only take enforcement action where the police identify that serious management failures have been a significant contributing factor to the incident.

If an employee is killed while driving for work and there is evidence that serious management failures have resulted in a 'gross breach of a relevant duty of care', the company or organisation could be prosecuted under the **Corporate Manslaughter and Corporate Homicide Act 2007**.

Like all other risks in the workplace, employers are required to manage the risks associated with driving at work, whether it is on a construction site or on public highways.

Plan

Assess the Risks

Driving (for or at work) should be treated the same as any other work activity, and risk assessment applied. The 'five-steps' principles should be followed when considering the driver, the vehicle and the journey.

TOPIC FOCUS

The HSE's approach to risk assessment:

1. Identify the hazards.
2. Identify the people who might be harmed and how.
3. Evaluate the risk and decide on precautions.
4. Record the significant findings and implement them.
5. Review and update as necessary.

The following issues need to be considered:

- **The Driver**
  - Are you satisfied that your drivers are competent and capable of driving in a way that is safe for themselves and other road users and pedestrians?
  - Do they have relevant experience, an up-to-date licence and a clean record?
  - Are you satisfied that drivers are properly trained for all types of vehicle they are likely to operate?
  - Have you ensured their fitness and health for the tasks, such as eyesight and mobility?
- **The Vehicle**
  - Is the vehicle suitable for the task and the driver?
  - Is the vehicle kept in a roadworthy condition (even site vehicles) and safe for all aspects of its operations?
  - Does the vehicle have all appropriate safety equipment (seatbelts, head restraints, etc.) and are they maintained? With site vehicles, this may include Falling Object Protective Systems (FOPS) and ROPS.
  - Do all vehicle operators have safety critical information to help them to reduce the risks (knowledge of vehicle requirements, such as tyre pressures, adjusting of lights, seat belts, head restraints, etc.)?
  - Are you satisfied that ergonomic issues do not pose a risk, such as seating position, and are drivers briefed on posture issues?
- **The Journey**
  - Are all routes planned thoroughly using safe routes appropriate for the vehicle?
  - Does route planning consider restrictions, such as narrow roads, height restrictions under overhead obstructions, weight limits on bridges, etc.?
  - Are there other hazards such as tunnels, railway crossings or ferry embarkation points, etc.?
  - Are work schedules realistic? Do they take account of hours of tiredness and sleep-related accidents (more likely between 2-6am and 2-4pm)? Do they try to avoid peak traffic flow ('rush-hours')?
  - Are you satisfied that enough time is allowed to carry out tasks and complete the journey safely such as the number of drops or collections to be made? Are overnight stays required to reduce journey times?
  - Can the distance be reduced or other modes of transport be used?
  - Are you satisfied that adequate consideration is given to the likelihood of adverse weather interrupting or prolonging the journey? Can the best routes be used to take account of this (major roads are more likely to be gritted in snow and ice, for instance)? High-sided vehicles will be most affected in high winds.

Policy

A safe driving policy should be in place and state the company's intentions to manage road safety. It should include what the management duties are - organising journeys (time/distance), driver training and maintaining the works vehicles.

The policy should take account of all legal requirements including those that restrict smoking in the workplace and the use of mobile phones.

Top Management Taking into Account Work-Related Driving

As with any health and safety matter, it is vital that work-related driving is taken seriously by senior management. They are ultimately responsible for the development and implementation of policies and procedures that control this issue.

They can set an example by planning business meetings at times when others will not face long journeys at busy times such as Friday afternoons.

Using public transport where possible and car sharing are some of the ways that driving for work can be reduced.

Roles and Responsibilities

A robust policy will clearly identify roles and responsibilities within the organisation. These duties may be agreed by management or they might be imposed by legislation. Various health and safety legislation imposes general duties on employers and employees which can be applied to the task of driving vehicles. The **Road Traffic Act 1991** and **Road Vehicle (Construction and Use) (Amendment) Regulations 2011** impose specific duties.

Do

Co-operations between Departments

The organisation and structure should allow co-operation between and across departments (or on site between contractors) with different responsibilities for road safety, for example, selecting and purchasing suitable vehicles and involving workers or their representatives in the process.

Adequate Systems in Place Including Maintenance Strategies

Employers must make sure they have adequate systems to allow them to manage work-related road safety effectively, for example, ensuring that vehicles are regularly inspected and serviced according to manufacturers' recommendations and legal requirements.

Communication and Consultation within the Workforce

The importance of good, clear communication and the benefits of consultation with the workforce was considered comprehensively in Element 2.

Employees must be made aware of the systems and strategies mentioned in the previous paragraph.

Provision of Adequate Instruction and Training

Employers should ensure that drivers have clear instructions about how to keep themselves safe while on the road. For example, the following information could be provided in the form of a handbook:

- How to carry out routine safety checks.
- How to correctly adjust safety equipment, e.g. seat belts and head restraints.
- What to do if the vehicle breaks down.

Make sure that drivers are aware of the height of their vehicle, both laden and empty. Drivers must be trained – giving priority to those at highest risk, e.g. those with high annual mileage, poor accident records, or those new to the job.

Check

Monitor Performance

Monitoring road safety performance ensures that the policy is working correctly. Employers need to check that all incidents and damages are reported, faults with vehicles are reported and promptly dealt with, and checks carried out (tachographs, etc.) of vehicle journey times and driver breaks.

Ensure All Workers Report Work-Related Road Incidents or Near Misses

A thorough policy must make it clear that all work-related road accidents, incidents and near misses must be reported to a line manager or other nominated manager.

Furthermore, it should state that all work-related road accidents, incidents and near misses will be investigated to establish how and why they occurred and to learn how to prevent them in the future.

To ensure the effectiveness of the policy, the organisation must have an accident reporting and investigation procedure that is rigorously implemented and must be followed by all staff and managers involved.

Act

Review Performance and Learn From Experience

The organisation should have effective arrangements in place to review the implementation and effectiveness of their policy on driver development, including:

- Reviewing adherence to assessment procedures and delivery of training.
- Reviewing the driving behaviour of its employees on roads.
- Identifying the extent of any shortfall in driver competence revealed by accident and incident investigation.
- Reviewing the impact of training on accident rates.

From the review process, lessons, both good and bad, must be learned and systems put in place to either maintain or improve on the current situation. One of the outcomes from the review should be a reappraisal of the current policy.

Regularly Update Policy

All policies can become out of date very quickly. The introduction of new legislation, and the introduction of new restrictions, such as congestion zones and following a serious accident, are some examples of factors that should prompt a review of the policy.

Work-Related Driving Control Measures

There are a range of general and specific measures which can be put in place to control the risks associated with driving at work. These include:

- Ensuring the policy is in place and understood by all drivers.
- Ensuring all management responsibilities are realised and achieved.
- Putting in place an organisational structure to manage the risks.

- Having appropriate safety systems in place.
- Monitoring the road safety performance.

Perhaps the most important consideration is to ask: can the hazard be eliminated? For example, a telephone call and video-conferencing could be instead of travelling to meetings.

If elimination is not possible, then consider:

- Using alternative transport (e.g. a train).
- Vehicle sharing to reduce the risk.
- Planning routes and delivery schedules realistically.
- Planning journey distance and times - avoiding excessive hours; staying overnight after long working day schedules instead of driving back.
- Scheduling stops for drop-offs and pick-ups of loads, passengers, etc.
- Ensuring vehicle maintenance is carried out and that defects are reported and dealt with.
- Ensuring driver competence - consider advanced driver training.
- Ensuring driver fitness and health appropriate for the tasks.
- Monitoring performance to ensure speed limits are complied with and adequate rest breaks are taken.
- Workers facing a long drive home at the end of the working day should be supported and encouraged to stay overnight and not drive when tired.

Safe Driver

Competence

Employers must check that their selection and training procedures ensure that drivers and other employees are capable of performing their work activities safely and responsibly.

The HSE suggests that employers consider:

- Checking that drivers possess the necessary licences or certificates for the vehicles they are authorised to drive e.g. lift trucks, shunt vehicles, site dumpers, etc.
- Checking the previous experience of drivers and assessing them to ensure that they are competent.
- Ensuring the provision of site-specific training on how to perform the job, and information about particular hazards, speed limits, the appropriate parking and loading areas, etc.
- Ensuring that there is a planned programme of refresher training for drivers and others to ensure their continued competence.

Certificates of completion of accredited training may be a requirement of employment for particular vehicles and certain vehicle operations. Employers should hold these records.

Operators' certificates should be examined before they undertake any driving tasks. The suitability of previous experience and training should be assessed to ensure it meets employers' current requirements. Where certification is not available, practical driving assessment should be undertaken.

Specific arrangements for ensuring understanding of, and skills in, operating in the particular conditions of a site may also be necessary, e.g. in relation to permit-to-work systems, where local codes of practice may be in operation during certain activities such as erecting/dismantling, heavy lifts or lifting operations, night shift work, flooding, excavation, temporary traffic light/traffic management systems, etc.

Validity of Driving Licence

Checking that employees hold valid driving licences is vital to ensure legal compliance and not doing so can have serious consequences. It's an offence to permit staff to drive without a licence but having a robust system in place can be a defence.

Regular driving licence checks should form a key part of any health and safety policy at a company where employees drive on business. This will uncover whether the member of staff is entitled to drive the type of vehicle that they are allocated and that they hold a current driving licence.

Simply asking drivers is not enough: a recent RAC Insurance survey found that 25% of motorists surveyed who already had penalty points did not inform any organisation when they received them, while only 13% would tell their employer if they incurred any.

Provision of Instruction

Provide information, instruction and training to maintain or improve employees' competence, particularly where changes in staff, equipment or procedures are planned. Construction sites and other types of site often have traffic routes for vehicles and pedestrians that change as work progresses. Employers should consider giving additional information to drivers.

Visiting drivers should report to the site office for any relevant instructions such as the workplace layout, which route to follow, and where to park, load and unload.

Information must be comprehensible. For some drivers may not speak English well, if at all, maps, diagrams and pictures can be used.

Fitness to Drive

An essential part of determining if a driver is 'qualified', is confirming the individual is fit to drive. Medical conditions and health issues such as sleep apnea, heart disease, diabetes, back strains and declining vision increasingly cause or contribute to motor vehicle incidents.

The law requires Heavy Goods Vehicles (HGV) drivers to be in reasonably good health in order to legally operate commercial vehicles on public roads. To that end, every new driver must undergo an HGV medical exam with accompanying paperwork that is submitted to the DVLA by the physician who performs the examination. Drivers are required to undergo a new medical exam every time their licence comes up for renewal.

Safe Vehicle

Fit for Purpose

The suitability of vehicles will depend on:

- The manufacturer's specification.
- The ground conditions the vehicle is required to operate on.
- Its stability, e.g. on muddy/soft ground, slopes/gradients, uneven surfaces.
- The effect of other work in close proximity, i.e. falling/rolling over, restraint systems, protection from falling objects, lifting or demolition work.
- Lights/warning devices.

- Adequate driver visibility, especially for reversing.
  - Driver protection from adverse weather conditions.

For example, a diesel-powered forklift truck would be unsuitable for use in a warehouse because of the toxic exhaust. A vehicle with large pneumatic tyres and high ground clearance would be suitable on rough ground.

Maintained in a Safe Condition

Employers are legally obliged to make sure that work equipment, including vehicles, is in good working order.

It is important that vehicles are maintained so that they remain mechanically sound.

Adequate Restraints for Securing Goods

It's very important to ensure that all loads are secure when transporting goods over any distance. It's impossible to estimate how much damage has been caused by unsecured loads falling from moving vehicles, but the vast majority of that damage could almost certainly have been prevented through the proper use of good-quality load restraint equipment. There are several load restraint systems on the market and they all have their own specialist application. Examples of the systems that are available are:

- Ratchet lashing straps.
- Cam buckle straps.
- Load binder kits.
- Ratcheting cargo bars.

Safe Journey

Planning Routes

Effective route planning helps drivers predict and prepare for hazardous aspects of being on the road, and therefore, plan to minimise risks.

For many drivers there is an over reliance on satellite technology. 'Sat Navs' are fine for the majority of journeys made in cars and small vans. For HGVs, buses and coaches they can cause considerable problems; they cannot identify low bridges, level crossings, tight corners, where roads narrow and restrictions imposed by local councils in residential areas.

Realistic Work Schedule

The EU imposes strict rules on drivers' rest breaks. In simple terms, the driver must rest for 45 minutes after 4.5 hours of driving.

Irrespective of the law, it is well known that tiredness is a major factor in road traffic accidents. Employers must be prepared to allow adequate time for the journey including overnight stops, if necessary, and taking account of local weather and traffic conditions.

Weather Conditions

Keeping an eye on the forecasts means that poor weather can be anticipated; ice or rain causing skidding or longer braking distances; sunlight blinding vision; rain or fog restricting visibility.

Consideration of Legal Driving Hours

The same EU rules that deal with rest breaks, also restrict the driving hours of HGV drivers. Briefly, they are:

- 9 hours daily driving limit, this can be increased to 10 hours twice a week.
- Maximum of 56 hours weekly driving limit.
- Maximum of 90 hours fortnightly driving limit.

STUDY QUESTIONS

10. Identify three factors associated with driving at work that can increase the risk of a traffic accident.
11. What four factors are used to ensure work-related driving is effectively managed?
12. Identify three main factors to consider when controlling work-related road driving.

(Suggested Answers are at the end.)

Summary

This element has dealt with some of the hazards and controls relevant to the movement of vehicles and plant.

In particular, this element has:

- Considered the hazards associated with the movement of people on construction sites, including striking by moving, flying or falling objects; collisions with moving vehicles; and striking against fixed or stationary objects.
- Explained the safe use of vehicles and plant within a construction environment, with particular reference to:
  - Hazards associated with loss of control; overturning; collisions with other vehicles or plant, pedestrians or fixed objects; and non-movement-related hazards.
  - Controls including:
    - Suitability of traffic routes.
    - Spillage.
    - Management of vehicle movements.
    - Environmental considerations, e.g. visibility, gradients, changes of level and surface conditions.
    - Maintenance and checking of traffic routes.
    - Means of segregating pedestrians and vehicles, and alternatives where separation is not practicable.
    - Protective measures for people and structures, e.g. barriers, signs and warning signals.
    - Site rules and signage.
    - Safe vehicles such as speed limits and particularly special rules relating to forklift trucks and parking.
    - Driver selection and training.
    - Assuring driver competence.
- Explained how to manage work-related driving using Plan, Do, Check, Act.
- Outlined work-related driving control measures including: safe site, safe vehicles and safe drivers.

Exam Skills

Question

Scenario

You have been tasked by the site manager to design a suitable traffic route for vehicles and to ensure the safety of pedestrians on site. The site is located next to a busy highway with a small river running along its rear edge.

Task: Pedestrian Safety

In order to design the layout of the traffic route, you decide to make a few notes covering the relevant aspects you need to consider in the design.

(Total: 10 marks)

Approaching the Question

Now think about the steps you would take to answer this question:

Step 1 The first step is to **read the scenario carefully.** Note that this is for your purpose so you need to consider the location of the site and possible hazards surrounding it. Think about the type of vehicles likely to be operating and the type of activities they will be involved in. Also give consideration to pedestrians and how interaction between vehicles and pedestrians could happen.

Step 2 Now look at the **task** - prepare some notes on considerations for vehicles and pedestrians when designing traffic routes.

Step 3 Next, consider the **marks** available. In this task, there are 10 marks so it is expected that around 5 or 6 different pieces of information should be provided. The task does not split the available marks – you will need to consider how the marks will be applied, so probably 2 marks for each piece of information.

Step 4 **Read the scenario and task again** to make sure you understand them and have a clear understanding of the hazards that are created by vehicle movements. (Re-read your study text if you need to.)

Step 5 The next stage is to **develop a plan** - there are various ways to do this. Identify key aspects of a vehicle route taking into account activities that will be conducted on site associated with vehicles. So, the answer plan will take the form of a bullet-pointed list that you need to develop into a full answer.

Suggested Answer Outline

Vehicle considerations:

- Suitability of traffic routes.
- Reversing/manoeuvring areas.
- Parking.
- Loading and unloading.
- Refuelling.
- Pedestrian routes.

Now have a go at the question yourself.

Example of How the Question Could be Answered

In order to ensure the safe movement of vehicles and pedestrians around construction sites, there are a number of aspects in their design that need to be considered.

First, the design of the vehicle route should be constructed of suitable material in order to withstand the weight of mobile plant and vehicles. It should be wide enough for the vehicles on site and where possible minimise steep slope and gradients.

Where possible, the need for reducing reversing and manoeuvring should be considered as this is a high-risk activity that can result in damage to structures, other vehicles and collision with pedestrians. Where possible, trying to eliminate reversing by having drive through practices and one-way systems. Where this is not possible, using banks persons who assist with the reversing operation. Manoeuvring areas need to be designated areas which should be large enough and well illuminated.

On site there will be a need for mobile plant and vehicles to be parked. These areas should be designated and located off the main traffic route system. Parking for staff and visitors' vehicles should be designated separately and located as close to the site entrance as possible.

Loading and unloading areas involve the movement of forklift trucks and therefore potential include additional hazards. These areas should be located close to material storage areas with minimum movement of people in the general area whilst loading and unloading is taking place.

Refuelling of on-site mobile plant should take place in a designated area on site with measures introduced to mitigate spillages of fuel, especially where there is a risk of contamination to water systems above and below ground. Provision for emergencies should also be made, in particular dealing with fire.

Where pedestrian routes come close to vehicle routes, then physical segregation using barriers need to be introduced with designated crossing points for pedestrians. In addition, the use of signage and additional illumination may be required.

Reasons for Poor Marks Achieved by Exam Candidates

- Not expanding the answer beyond a few words as opposed to giving a sentence of explanation.
- Misinterpreting the question by identifying vehicle and mobile plant hazards and control measures.
- Focussing on two or three areas and going into too much detail opposed to going wider covering all relevant points.

Element 7

Working at Height

Learning Objectives

Once you've studied this element, you should be able to:

1. Explain the hazards of working at height and outline the precautions necessary to control them.
2. Explain safe working practices for access equipment and roof work.
3. Outline the control measures necessary to protect other persons not involved in the work at height.

Contents

Working at Height Hazards and Controls

IN THIS SECTION...

- The hazards and factors affecting the risk from working at height include the vertical distance of a fall, fragile roofs, deteriorating materials, unprotected edges, unstable or poorly maintained access equipment, adverse weather conditions, and falling materials.
- Work at height should be avoided where possible, or engineering methods used to prevent falls and falling materials, including proper planning and supervision of work, and avoiding work in adverse weather conditions.
- Emergency rescue plans should be in place to facilitate the rescue of workers.
- Training and instruction are required to minimise the distance and consequences of a fall.

The Risks of Working at Height

Vertical Distance

The vertical distance a person or object can fall will have a direct bearing on the severity of injury or damage it can cause. Where work at height cannot be avoided, the first considerations will be to reduce the distance a person can fall.

Roofs

Fragile Roofs

Roof work includes construction and maintenance of roofs, e.g. replacing tiles, gutter cleaning, chimney repairs and re-pointing. Many accidents occur during small jobs and maintenance work. In addition, maintenance work is often undertaken in a hurry, e.g. storm damage repairs.

A fragile roof is any roofing structure which is not:

- Specifically designed to withstand the weight of a person or a load they may be carrying.
- Strong enough to withstand adverse weather conditions.

Roofing materials such as cement, asbestos, glass, reinforced plastics and light tongue-and-groove wood covered with roofing felt are all liable to collapse under the weight of a worker.

All fragile roofs except glass should have a large warning notice displayed on them. However, remember that with a multinational workforce, a sign in English does not necessarily provide sufficient warning and must be backed up by verbal warnings.

The safe method of working on fragile roofs is by the use of roof ladders or crawling boards. These are laid across the surface, supported by underlying and load-bearing roof members, and distribute the load of the worker over a wider area, enabling the roof structure to sustain the load safely. Roof ladders also provide a good foot-and-hand hold for the worker.

Fall arrest equipment should be worn when shown to be required by a risk assessment.

Roof lights are commonly clear sheets built into a roof to allow light through, and are made of fragile materials (such as plastics) and not always easy to see - they may be covered with dirt and grime, moss or algae, or maybe painted over. They are often difficult to visually distinguish from the surrounding roof materials in bright or very dim light. If they are not covered, guarded or fitted with toe boards, it can be easy to fall through and suffer serious (or fatal) injury.

Sloping (Pitched) Roofs

These are roofs with a pitch greater than 10°. Falls from the edges of sloping roofs generally cause serious injury even when the eaves are low, as on a bungalow. If the person has slipped down the roof from the ridge, considerable acceleration can be generated which tends to project them from the eaves, adding to the force of impact with the ground and hence to the seriousness of the injuries sustained.

Flat Roofs

Work on flat roofs can lead to falls (usually from unprotected edges). These commonly occur:

- From a completed roof.
- From a roof where work is being completed.
- Through openings or gaps (voids).
- Through fragile materials making up part of the roof, e.g. roof lights.

Deterioration of Materials

The condition of the structure on which people are working should be sound. However, materials deteriorate over time, particularly when exposed to the effects of the weather (heat as well as cold and water) and attack by animals and insects, etc.

Unsound materials represent a hazard in two ways:

- The danger of the material breaking when a person puts their weight on it and falling through the surface.
- The danger of materials breaking off and falling to hit people or structures at lower levels. Where they hit lower structures, this may cause a collapse.

It may not always be evident that deterioration has occurred until it is too late, so care needs to be taken to ensure that footholds are sound and secure.

Unprotected Edges

Where the edges of surfaces on which people are working are open, the risk of falls or falling objects is greatly increased. This applies to roofs, elevated walkways, scaffolding and access platforms, etc.

We'll consider the use of guardrails, toe boards, fencing and other protective boarding required to prevent such accidents later.

Unstable or Poorly Maintained Access Equipment

Access equipment includes scaffolding, towers, platforms and ladders. There are inherent risks in using such equipment, but these are compounded if the equipment is unstable, unsecured and not properly maintained.

Any access equipment that is not positioned correctly, is poorly constructed or poorly secured will be unstable. Conditions such as overloading of the equipment, high winds or overreaching can then cause a catastrophic collapse or topple.

Equipment not properly maintained can suffer damage from rust, physical damage, broken or damaged treads and fittings, and platform boards that rot through the timber.

Weather Conditions

The weather can increase the risks associated with working at height:

- Rain or freezing conditions can increase the risk of slipping.
- High winds can make access equipment unstable and blow loose materials off (and in extreme conditions, even workers).
- Cold conditions can cause loss of manual dexterity and can lead to an increase in muscle injuries.

Falling Materials

Objects falling from a height are capable of causing considerable damage to both people and other materials that they hit. The objects themselves may be:

- Loose structural material, e.g. tiles, bricks and timber.
- Waste materials, e.g. stone chippings or off-cuts of wood.
- Equipment or tools which are dropped.

Circumstances which contribute to the likelihood of falling materials include the following:

- Deterioration of structures causing crumbling brickwork or loose tiles.
- Bad storage of materials, e.g. at the edges of scaffold platforms, or in unstable stacks.
- Poor housekeeping leading to accumulations of waste and loose materials.
- Gaps in platform surfaces or between access platforms and walls.
- Open, unprotected edges.
- Incorrect methods of getting materials, equipment or tools from ground level to the working area.
- Incorrect methods of getting materials down to ground level, e.g. throwing.

Approach to Working Safely at Height

All work at height should be assessed. The best way to manage the risks inherent with working at height is to eliminate the need to work at height altogether. This can be achieved by:

- Modifying a work process, e.g. cleaning windows from the ground by pole cleaning rather than off ladders.
- Good design, e.g. erecting guardrails or steelwork at ground level and then craning the steel and guardrails into place.

However, in most instances in the construction industry, avoidance will not be possible and control measures for working at height will be required.

TOPIC FOCUS

The **Work at Height** (**WAH**) **Regulations 2005** impose a **risk prevention hierarchy**:

- **Avoid** work at height.
- Use work equipment or other measures to **prevent falls** where work at height cannot be avoided.
- Use work equipment or other measures to **minimise the distance and consequences** of a fall where the risk of a fall cannot be eliminated.

Avoid Work at Height

Not doing work at height is the best solution to the problem - do as much work as possible at ground level. Assembling components which can then be lifted into place is one way - using long-handled or telescopic tools (e.g. for cleaning windows) offers an alternative. Prefabrication or ground-level fabrication offers a way of reducing work at height e.g. assembly of roof trusses.

Prevent a Fall From Occurring

- **Using a Known Place**

  Using a place that is known to be safe such as a solid roof with fixed guardrails.

- **Use of Suitable Equipment**

  Provide a safe working platform such as a Mobile Elevating Work Platform (MEWP) or scaffold with guardrails, fences, toe boards, etc. that are strong enough to prevent a fall.

- **Work Restraint Systems**

  If the previous options are not possible and a worker can approach an unprotected edge, provide equipment which will restrain or prevent this from happening.

- **Minimise Distance and/or Consequence of a Fall**

  Where work at height cannot be avoided, employers must minimise the distance a person can fall, and if not, the consequences of a fall.

  Soft landing systems such as safety nets and airbags that are fixed close to the level of work will give an element of protection to anyone who falls whereas fall restraint, fall arrest will only protect the person using it - assuming it is being used correctly, if at all. Priority must be given to those methods which protect more than one person.

Main Precautions Necessary to Prevent Falls and Falling Materials

Proper Planning and Supervision of Work

Work at height should be planned in advance, with careful consideration given to the selection and use of work equipment. A safe system of work should be set up which takes account of:

- Levels of supervision of workers required, e.g. fall arrest equipment will require a higher level of supervision than work on a mobile scaffold tower.
- Weather conditions, e.g. carrying out maintenance on an icy roof or working in rainy conditions on a slippery surface.
- Emergency or rescue arrangements that may be required, e.g. if workers fall while using a fall arrest system. It is not acceptable just to rely on the emergency services; this needs to be covered in the risk assessment and planned beforehand.

Objects can be prevented from falling by:

- Not stacking materials near edges, and particularly unprotected edges.
- Close boarding of working platforms to minimise the gaps between scaffold boards, or placing sheeting over the boards so that material cannot fall through.

- Avoiding carrying materials up or down ladders, etc. by using hoists and chutes to move materials.
- Using physical safeguards such as toe boards and brick guards.
- Where a risk remains, using physical safeguards to prevent falling objects hitting people below, e.g. debris netting, fans (wooden shielding angled to catch debris) and covered walkways (tunnels).

Avoiding Working in Adverse Weather Conditions

If adverse weather such as icy, rainy or windy conditions greatly increases the risk of working at height (e.g. carrying a wide roof sheet in high wind), the work should be postponed until conditions are satisfactory. Getting a daily weather forecast is a suitable precaution (five-day weather forecasts can also be used). Wind speeds and temperatures at height can be considerably different to measurements taken at ground level. Winds can funnel and eddy especially where there are other high-rise buildings.

This stipulation does not apply for emergency services acting in the event of an emergency.

Emergency Rescue

Emergency procedures should be in place to cover reasonably foreseeable circumstances, e.g. stuck access equipment or deployed fall arrest, so that people can be rescued.

Methods selected need to be proportionate to the risk:

- There should be no reliance on the emergency services.
- Simple systems may suffice (e.g. putting a ladder up to a net and allowing a fallen person to descend).
- More detailed systems may be required (e.g. the use of other work equipment such as MEWPs or proprietary rescue systems such as a rescue harness, pulleys and clamps).

Employers must ensure that those involved are trained in the procedures and that the equipment required is available.

Provision of Training, Instruction and Other Measures

In all cases, operatives must receive adequate training and instruction to appreciate the dangers of working at height, as well as how to service and use the equipment, and the correct methods of use.

STUDY QUESTIONS

1. What is the hierarchy of risk prevention measures for working at height?
2. What are the safe methods of working on a fragile roof?
3. Where can falls occur from a flat roof?
4. What measures should be taken to prevent materials falling from a height?

(Suggested Answers are at the end.)

Safe Working Practices for Access Equipment and Roof Work

IN THIS SECTION...

- Scaffolding provides access to work at height and includes independent tied, putlog, fan, cantilevered and mobile tower scaffolds.
- Safety features of scaffolding include sole boards and base plates for ground support; toe boards and guardrails to prevent falls; and boarding, brick guards and debris netting to retain materials, equipment and debris.
- Scaffold erectors must be properly trained and qualified to erect, maintain and dismantle scaffolding.
- Safe access onto and off scaffolding has to be provided, and loading platforms must be suitable and strong enough to support the work.
- Various equipment and personnel hoists are available to move people and materials on and off scaffolding and building structures.
- Scaffolding must be stable, and its structural integrity must not be affected by loading of people, equipment and materials or adverse weather conditions. It must be protected from impact by vehicles.
- Other access equipment includes ladders, stepladders, trestles and staging. Where roof work is next to a gap or edge, then further leading edge protection is required.
- MEWPs offer a portable alternative to full scaffolding in some situations.
- Other techniques include boatswain's chairs, cradles and rope access systems.
- Fall arrest systems include harnesses, safety nets, soft landing systems and crash decks. Rescue and emergency procedures should be in place when using these techniques.
- Safe roof work requires safe access, edge and leading edge protection, and crawling boards on fragile materials and fall arrest equipment.

Scaffolding

Design Features of Different Types of Scaffold

Scaffolding is made up of the following basic components:

Scaffolding

- **Standards** - uprights or vertical tubes used to support the load to the ground.
- **Ledgers** - short horizontal tubes tying the structure together longitudinally, usually running parallel to the face of the building.
- **Transoms** - short horizontal tubes spanning across ledgers normally at right angles (90°) to the face of the building. They may also be used to support a working platform.
- **Bracing** - diagonal tubes to give the structure its rigidity.
- **Base plates** - small square metal plates that the standards (uprights) rest on to prevent them sinking into the ground.
- **Sole boards** - large pieces of timber put under the base plates to spread the load over a wider surface area when the scaffold is erected on soft ground.

- **Work platform** - the fully boarded work area surface.
- **Guardrails** - fixed to the standards (uprights) to fully enclose the work platform.
- **Toe boards** - fixed to the standards (uprights) to provide a lip at the platform edges.

Independent Tied Scaffolds

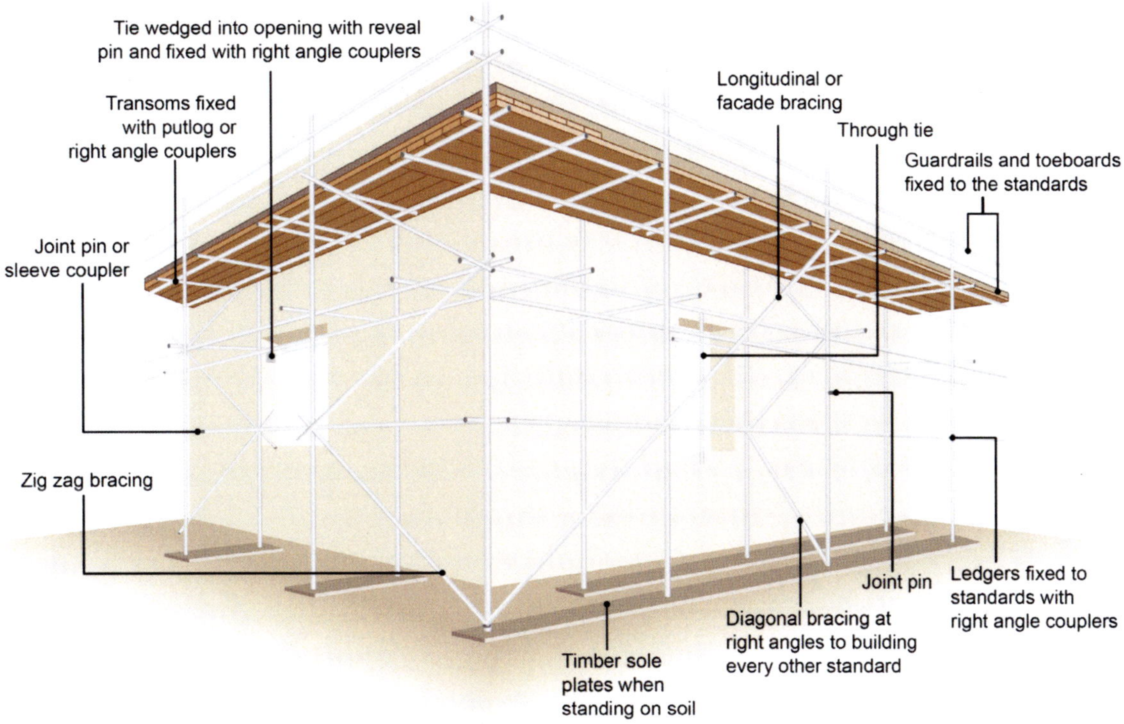

Independent tied scaffold

An independent tied scaffold is designed to carry its own weight and the full load of all materials and workers on the platform. It must be tied to the building where it is sited, to give stability and prevent movement.

As the total weight of the structure is supported by the ground, it is very important that the ground conditions are suitable to cope with the load. Base plates and sole boards may be used to spread the weight over a large surface area.

There are a number of ways in which the scaffold can be tied to the building to prevent movement:

- **Anchor bolts** - where one end of a metal bolt is screwed into the wall of the building and the other end is attached to the scaffold tubing.
- **Through ties** - where a scaffold tube extends into the building through an opening (e.g. a doorway or window frame). The end of the tube inside the building is coupled to another tube bridging across the opening.

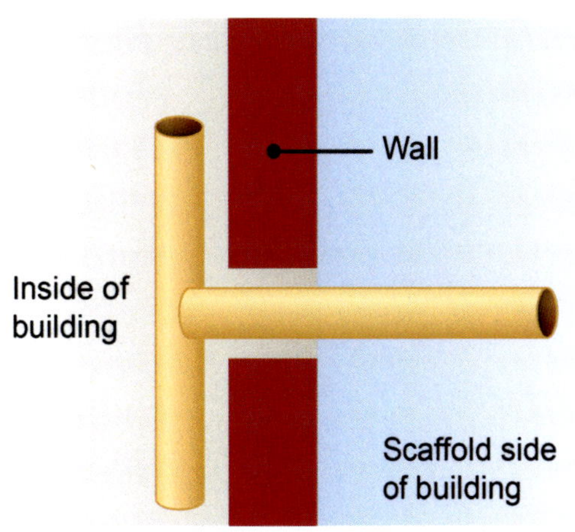

Through tie (as seen from above)

- **Reveal ties** - where a scaffold tube is coupled to a reveal pin that is wedged tightly across an opening such as a window reveal.

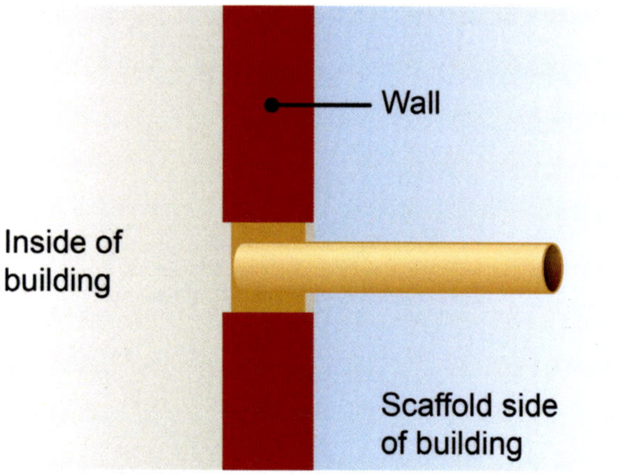

Plan view of a reveal tie

- **Rakers** - supports which push the scaffold onto the building. This system takes up space and may not be suitable in urban environments.

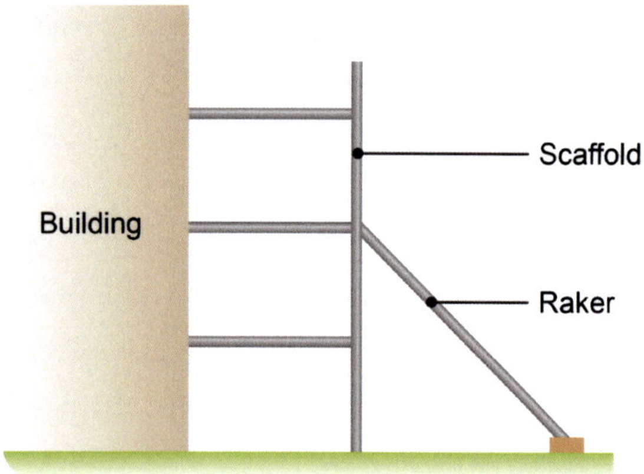

Side view showing rakers

Bracing is another important component of the scaffold and is used to stiffen the scaffold framework to prevent collapse.

The basic forms of bracing are:

- **Facade bracing** - scaffold tubes running diagonally across the face of the scaffold. This is also known as sway and longitudinal bracing.
- **Ledger bracing** - scaffold tubes running diagonally within the framework. Alternate pairs of standards should have ledger bracing.

TOPIC FOCUS

Factors that might cause the collapse of an independent tied scaffold:

- Overloaded work platform.
- Scaffold built on soft ground without sole boards.
- Scaffold not adequately tied to the building.
- Insufficient bracing incorporated into the scaffold.
- Standards not upright.
- Standards bent, buckled or heavily corroded.
- High winds.
- Incorrect couplers used to join tubes together.
- Scaffold struck by mobile plant.
- Scaffold erected by incompetent workers.
- Scaffold not properly inspected.

Putlog Scaffold

A putlog is a tube spanning from the ledger to the wall of a building on a putlog scaffold. It is fixed into the brickwork by a specially formed flattened end (spade end) which may be constructed on a scaffold tube, or by using a detachable fitting (putlog end) with the spade end of the putlog fully home (75mm) in the brickwork (bed joint).

Putlog scaffold showing spade end

Putlog end

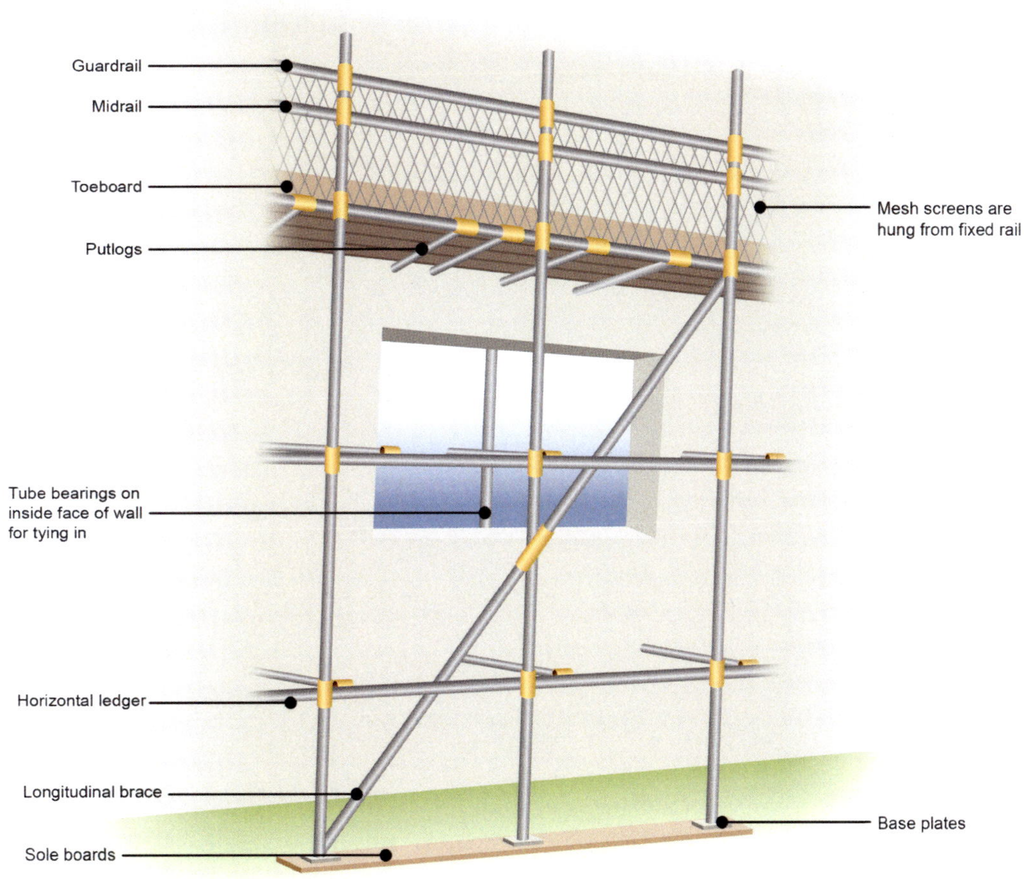

Putlog scaffold

TOPIC FOCUS

Scaffold Inspections

These are carried out:

- After erection and before first use.
- Every seven days.
- After any adverse event, e.g. bad weather, impact from vehicle, or large object falling into it.

The inspector should look for:

- Adequate foundation - level, firm ground (base plates should be central on sole boards).
- Platform - fully boarded, three to five boards wide (depends on use).
- Scaffold boards - each to have at least three supports and be tied down (overhang not more than 150mm).
- All tubes and fittings are in good condition.
- Vertical supports - no more than 2 to 2.5m apart.
- Bracing - along diagonals at the faces and at right angles.
- Ties - into the building at least every 4m vertically and 6m horizontally.
- Guardrails and toe boards - secure along outer edges and ends.
- Toe boards - minimum of 150mm high.
- Guardrails - 950mm above the platform; double guardrail at 460mm.
- Appropriate access.
- Ladders - good condition (correctly sited, tied, sufficient length and number, etc.).
- Loading - do not overload; stack any material correctly and securely.

Additionally for a **putlog scaffold**, checks are needed on the spacing of putlogs and to see that the spade end of the putlog is fully home (75mm) in the brickwork (bed joint).

Fan Protection

Fans or protected walkways are often used to protect members of the public from falling materials and debris from work carried out in busy city centres, but can equally be used on construction or refurbishment projects.

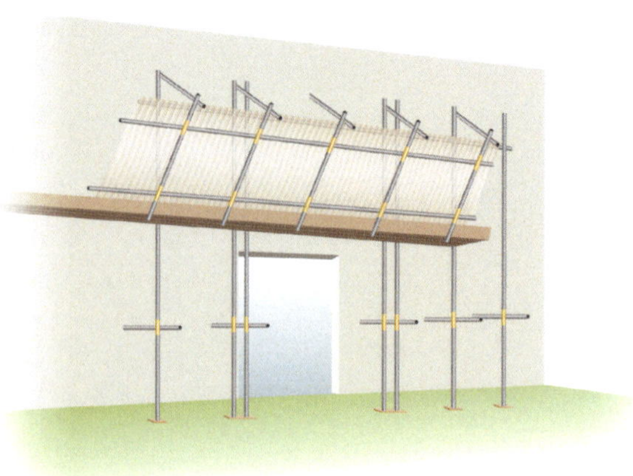

Fan protection

Cantilevered Scaffold

A cantilever scaffold is an independent scaffold entirely supported by the building. The scaffold has a framework inside the building, wedged between floors and ceilings. The cantilever tubes are fixed to this frame and supported. Rakers bearing on sills or ledges on the face of the building support the outer ends of the tubes. Buildings should be checked to ensure they can support the intended loads.

Mobile Tower Scaffolds

Commonly used for painting and simple maintenance jobs inside and outside buildings, they are only light-duty scaffolds. They have one working platform at the top, accessible by fitted internal ladders. They can use scaffold tubes, but most use proprietary-made structures. The assembled tower is mounted on wheels and can be easily moved. They are to be built to manufacturers' instructions by competent persons.

Worker enters top of tower scaffold via trapdoors using internally fixed ladders; tower has outriggers attached

The safe height of a mobile tower is determined by a complex calculation taken from BS EN 1004 which takes into account many factors. Manufacturer's instructions should be used to determine maximum heights and if outriggers are required. If they are required, they should be fitted at the earliest opportunity. Maximum indoor heights of towers should not exceed 12m while the maximum outdoor height should not exceed 8m. If the average wind speed reaches 17mph (Beaufort 4, 27km/h or 7.6m/s), you should cease work and consider dismantling the tower. Alternatively, the tower should be tied to a rigid structure. The tower should be dismantled and safely stored if wind speeds are above 40 mph.

Duties regarding edge protection, overloading and inspections are the same as for other scaffolds/working platforms.

Main hazards associated with mobile tower scaffolds:

- Falls from the work platform.
- Objects falling from the platform.
- Collapse of the structure.
- Overturn (toppling) of the structure.
- Unintended movement of the wheels.
- Contact with live overhead cables or other services.

TOPIC FOCUS

Safety precautions for safe use of mobile tower scaffolds:

- Guardrail fitted to the work platform.
- Tower must not be overloaded.
- Wheels should be locked when tower is in use.
- Tower must be sited on firm, level ground.
- Never move towers with people or materials on them.
- Avoid overhead services during use and movement.
- Use outriggers where necessary to increase stability.
- Never climb up the outside of a tower.
- Towers must be assembled by trained workers.

Scaffolding Hazards

Key hazards associated with the use of scaffolds:

- Falls from the work platform.
- Objects falling from the platform.
- Collapse of the structure.

Scaffolds should be erected by trained workers and inspected by competent persons.

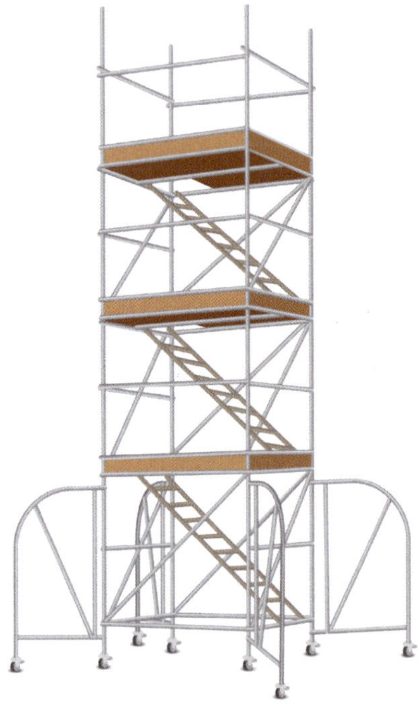

Mobile scaffold tower

Safety Features

Sole Boards

The sole board is a strong timber plank, wider than a base plate and long enough to be positioned under two base plates. Sole boards spread the load for the scaffold base. They must be used on soft ground (it is good safety practice to use them for all scaffolds).

Base Plates

A base plate is a flat, square steel plate with a locating pin that must be inserted into the bottom of a standard to provide a bearing surface for load distribution.

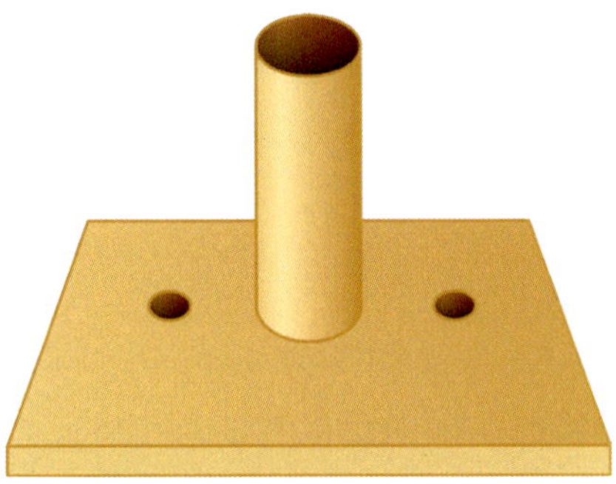

Base plate

Toe Boards

Safeguards must be provided for all unprotected edges on scaffolds and platforms.

Toe boards are usually scaffold planks laid on their edge at right angles (90°) to the working platform. They are laid at the outer edges and ends of the working platform (sometimes the inner edge - the one nearest the building - may also require them). Toe boards prevent small objects, e.g. rubble and tools, from being kicked off the platform.

A toe board must extend to at least 150mm above the working platform.

Guardrails

Guardrails will prevent people from falling. Detailed requirements are:

- Top guardrail to be at least 950mm above the working platform.
- Intermediate rails so no unprotected gap of more than 470mm between toe board and guardrail.

These arrangements are used on a scaffold platform to prevent falls from a sloping roof. The same arrangements apply in preventing falls in respect of flat roofs, and also other forms of edge protection such as mobile tower scaffolds, mobile elevating work platforms and elevated walkways.

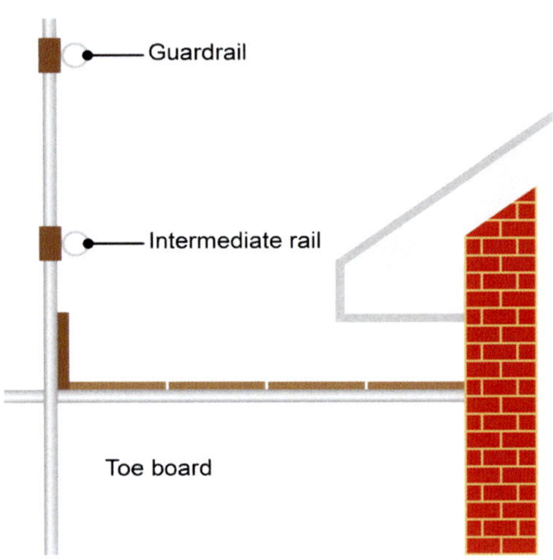

Guardrails/toe boards

Boarding

An intermediate guardrail does not have to be used, although there must be no vertical gaps exceeding 470mm. **Boarding**, consisting of several toe boards stacked one on top of the other, could be used instead, or a brick guard hung from the main guardrail and attached to the platform. Providing this could support/contain a person in the event of a slip or fall (in which case it might be possible to do away with the toe board).

Gaps are allowed in edge protection, provided they are the minimum required for reasonable access. Lateral gaps in guardrails are only permitted at a point of access to a ladder or stairway where a gap is necessary.

As an alternative, barriers or fencing can be erected around roof edges using scaffolding tubes and boards. These must extend at least 400mm above the roof surface and the upper guardrail must be at least 910mm above the roof surface.

Brick Guards

If material is stored above toe board height then additional boards, brick guards (mesh panels) or similar mesh covering are used to prevent material from falling and have a secondary function of helping prevent people from falling as well.

Debris Netting

Safety nets can also be used to prevent waste material or debris from falling by using them in the shape of a fan to protect those below. In order to achieve this, overlay nets (debris nets) are laid over the safety nets. Care must be taken to ensure that both the size and spacing of the overlay net are in line with the type of material falling on it, e.g. 12-19mm. As with safety nets, overlay nets need to be attached to the supporting framework with tie cords, hooks, rings or thimbles and should be double the strength of the safety net. Such nets should not be allowed to accumulate debris above a certain weight as specially designed nets are required for this purpose. The nets should be cleaned, inspected and stored in accordance with the precautions outlined for safety nets (see later) by a competent person.

Use of a brickguard

Requirements for Scaffold Erectors

Scaffold erection is a skilled and potentially dangerous occupation, and requires the necessary capability and fitness, skill, experience, knowledge and competence to complete scaffolding operations safely.

Anyone engaged in work at height and the use of work equipment for work at height, including organisation, planning and supervision, must be competent to do so. If being trained, they are to be supervised by a competent person.

Scaffold erectors need to be trained in the following:

- Safe systems of work.
- Risks associated with scaffolding and with any particular work involved.
- Assessing site conditions.
- Use of PPE.
- Erecting, adapting and dismantling different types of scaffolding.

- Setting up hoists or lifting appliances.
- Operating mobile access platforms.
- Rigging and inspection of safety nets.
- Application of the **Construction (Design and Management) Regulations 2015** and the **Work at Height Regulations 2005** in the operations above.

Scaffolders are classified as basic (one year of experience) or advanced (two years' experience including one year as a basic scaffolder). New entrants are called trainee scaffolders.

Means of Access

Stair towers are now readily available as a safe means of access to scaffolding. Stair towers can speed up construction activities because workers can more readily carry materials up them. Stair towers must be properly tied into the structure or to the scaffold. Access onto scaffolding by the use of ladders should be assessed by a risk assessment. If the ladder is over 9m in height, then a safe landing area or rest platform should be provided at suitable intervals. Access to scaffolding for people and materials might also be gained by using hoists (discussed later in this section) and mobile elevating platforms.

Design of Loading Platforms

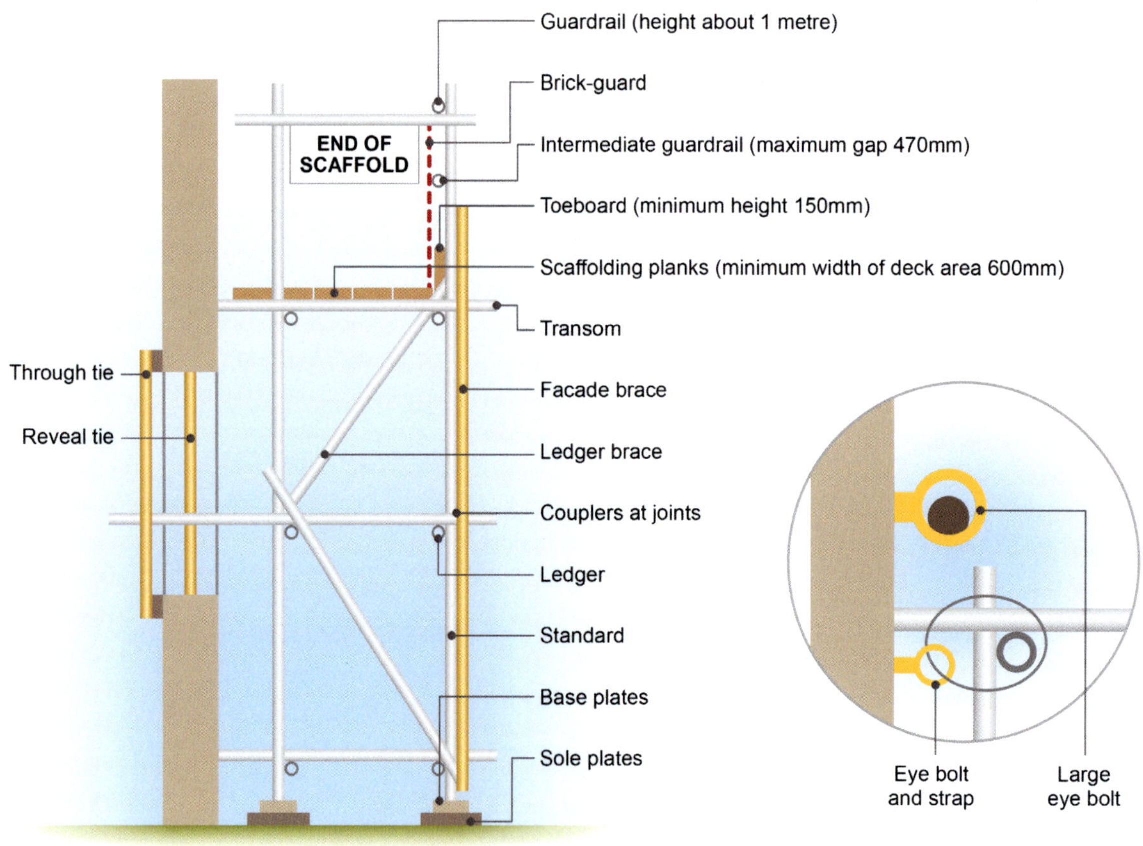

Independent tied scaffold with working platform

Loading/working platforms and any supporting structures must not be overloaded so as to collapse, or be deformed in any way so that they are unsafe to use or give rise to any danger. They should be designed, erected and maintained by a competent person.

Working platforms and decking should be closely boarded to their full width. If boards overlap, bevelled pieces should be used to avoid tripping hazards. The platform must be wide enough for the work being done:

- **Three boards wide** (600mm) - people doing light work without materials (e.g. inspection or painting).
- **Four boards wide** (800mm) - person and materials; jobs like glazing, plastering.
- **Five boards wide** (1,050mm) - a general purpose work platform (general building work and bricklaying). This can carry trestles or similar platforms for work at higher levels.
- **Six boards wide** (1,300mm) - specially designed - for dressing or roughly shaping stone.
- **Seven boards wide** (1,500mm) - heavy duty decking, for dressing stone and to support stone-dressing platforms.

A clear passage 440mm wide should be kept on platforms, and the platform should be no further than 300mm from a wall. Platforms should extend 600mm beyond the end of a wall.

Scaffold Hoists (People and Material)

Manual hoists are used to raise lightweight equipment and materials to a place of work, e.g. block and tackle, hoists, wire ropes, chains or slings.

Different types of **mechanical hoist**, e.g. vertical, swivel and beam, are available to transport materials and people.

Goods hoist

All scaffold hoists must be:

- In good repair and well maintained.
- Marked with a safe working load.
- Positioned on stable ground according to the manufacturer's instructions by a competent person.
- Protected at both top and bottom to prevent falls of materials or persons.
- Clearly marked in respect of the control functions.
- Supervised at all times when being used.

Hazards associated with the use of hoists include:

- Falling down the hoistway:
  - From a landing level, e.g. scaffolding.
  - From a platform or cage.
  - With the platform/cage due to operator error while loading or unloading is taking place, or due to gear failure.
- Being struck by:
  - The moving hoist or cage.
  - Falling materials from the hoist.
  - Landing levels or parts of the enclosure while trying to ride on material/goods hoists.

The safe operation of hoists depends on:

- Routine testing and examination by competent persons.
- Training of operatives.
- Security of the loads.
- Adequate control measures:
  - No riding on a goods hoist.
  - No gaps at the edge of the hoist platform.
  - Enclosure of the hoistway.
  - Hoist gates at all levels and fencing where people may fall down.
  - Landing gates kept closed during loading/unloading from the ground.
- Clear operator visibility to all levels of the installation.

Two types of **passenger hoist** are commonly used:

- Surrounding tower hoist.
- Cantilevered hoist.

Both must be adequately tied (ideally to the permanent structure) at suitable intervals, and only operated within their Safe Working Load (SWL):

- The surrounding tower type has the cage running in the centre of the tower.
- The cantilever type has either one or two cages running on either side of the tower.

These systems have a number of **safety devices**, including:

- Over-run devices.
- High landing gates (minimum two metres).
- Hoist arrester devices.
- Deadman's handle.
- Cage fully enclosed (and enclosure a minimum of 2m in height).
- Hoist tied into the building.
- Hoist operated from one position.
- Cage interlocks on gates.
- Landing gate interlocks.
- SWL on hoist.

Ensuring Stability

The following ensure stability:

- Requirements for scaffold structures:
  - Rigid.
  - Stable.

  - Strong enough to support working platforms and any load that might be placed on them, e.g. bricks, tools, equipment.
  - Erected so that they cannot be displaced and so give rise to any danger.
- Structure must remain stable:
  - Even when being altered or modified.
  - During dismantling operations.
  - In high winds or inclement weather, e.g. slippery, icy surfaces.

  Sheeting or netting may also increase the load on structures during high wind conditions.
- Care is required to avoid impact from vehicles and possible collapse of the structure.
- There should be proper segregation, marking and lighting of structures.

It is worth noting here that the collapse of a structure such as a scaffold would be a dangerous occurrence, and therefore reportable under **RIDDOR**.

Effects of Materials

An independent tied scaffold is designed to carry its own weight and the full load of all materials and workers on it. The total load is supported by the ground the scaffold is built on. The scaffold is not totally independent and must be tied to the building where it is sited, to prevent movement away from or towards the building.

As the ground supports the total weight of the scaffold when in use, the ground structure must be suitable to cope with the load. Base plates and sole plates will help to spread the load. Overloading the scaffold with excess materials is liable to make the structure unstable.

Weather

The wind strength and adverse weather conditions, e.g. rain, ice or snow, must be taken into account. At ground level, wind may be reduced by the protection given by buildings; at roof level a worker may be exposed to its full force. Footholds at heights are generally not as good as at ground level and balance may be affected. Carrying large items of material, e.g. a roofing sheet or lengths of wide timber, will exaggerate the problem by acting as a 'sail' in the wind. Workers have been blown off a roof while handling large items of materials.

When deciding whether to suspend work it is important to consider:

- The wind speed.
- Measures already taken to prevent falls.
- The positioned height of the roof or work equipment and size of materials being handled.

Sheeting

Two main types of sheeting are commonly employed on scaffolds:

- **Debris netting** - comprised of an open meshwork which reduces the risk of falling debris outside the work area. While not reducing the effects of adverse weather, it may reduce wind loading on the scaffold.
- **Scaffold sheeting** - made of tough reinforced polythene material, often with eyelets to connect sheets together. It has a high tensile strength and is tear and puncture resistant. Sheeting material with strong overlapping joints will reduce the effects of adverse weather conditions within the scaffold, and the effects of very strong winds.

Sheeting and netting must not be attached to the scaffold unless the scaffold has been designed to resist the additional load caused by winds.

Protection from Impact of Vehicles

Protection from vehicles can be ensured by the following measures:

- On separated sites, vehicle access areas should be segregated and fenced/coned-off to prevent construction site vehicles impacting with scaffolds.
- In public areas, traffic control measures may need to be used to prevent vehicle impact.

Inspection Requirements

Scaffolding should always be inspected by a competent and appropriately trained individual. To meet the requirements of the **Work at Height Regulations 2005** scaffolding should be inspected:

- After installation/prior to being used.
- At least weekly thereafter.
- Following any circumstances which could jeopardise the safety of the installation such as adverse weather conditions. So even if a scaffolding structure was inspected just the day before, it should be inspected again if, for example, there were high winds overnight or reports of events such as an earthquake in the region.

It is the responsibility of the hirer/user to ensure that scaffolding has been inspected in line with the regulations.

A written report should be completed for each inspection and retained on site until work is completed. Once construction work is complete, reports should then be kept at an office location for a further three months.

The following information should be included in every scaffolding inspection report:

- Name and address of the person for whom the inspection was carried out.
- The name and position of the person carrying out the inspection.
- The location and the date and time.
- A description of the place of work or where the work equipment was inspected.
- Details of defects or anything identified during the inspection that could pose a risk to the health or safety of any person.
- Any action taken to remedy this.
- Details of any further action considered necessary.

Use of Equipment

Ladders

Ladders should only be used for low risk activities of short duration (e.g. inspection or painting). Three points of contact should be maintained at all times on ladders. They are also often used to provide access and egress, e.g. into excavations and on and off scaffolds.

Ladders must be regularly inspected and maintained, and should be built to a suitable standard (i.e. **BS EN 131**: maximum static load of 150 Kg).

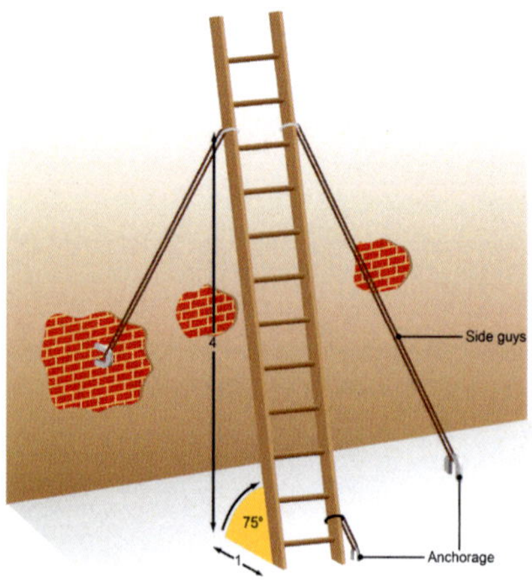

Positioning and securing a ladder

Main hazards associated with the use of ladders are:

- Not being tied or not resting on firm ground, which may cause the ladder to tip or topple sideways.
- Poor storage and maintenance allowing the ladder to rot or warp, which may cause the rungs to break.
- Objects falling from height.
- Contact with live overheads.

Safety precautions for use of ladders are:

- Do not handle or position ladders near live overheads.
- Position on a solid, flat base so the feet do not sink into the ground. Weight should only be supported on the styles, never on the rungs. Tie where necessary.
- Ideal ladder angle is 75° to the horizontal or at a ratio of 1:4 distance away from the wall to height (1 out: 4 up).
- Top of the ladder must rest against a solid support (e.g. not plastic guttering).
- Ideally the ladder should be secured at the top. If this is not possible, then guy ropes should be attached and secured to firm supports. If this is not possible, the ladder should be supported by use of a proprietary ladder stabiliser or 'footed' by someone standing with one foot on the bottom rung while holding both styles.
- Top of the ladder should extend far enough above the level of the working position or the platform onto which it gives access to provide a safe handhold. The stepping-off point should be safe and clear.

Ladder used to gain access to a scaffold platform. Note how the ladder extends well above the stepping-off point and how it has been secured to the scaffold

- Only one person should climb on the ladder at any one time.
- Nothing should be carried in the hands going up and down the ladder, leaving them free to hold both styles.
- Never over-reach out to the sides of the ladder.
- Wooden ladders should not be painted - this can hide defects.

Extension ladders - those that slide up in sections - must have sufficient strength at the overlap. The guide is:

Closed Ladder Length	Overlap
Less than 5m	1.5 rungs
5m to 6m	2.5 rungs
Over 6m	3.5 rungs

Stepladders

Hazards associated with the use of stepladders:

- Falls from height:
  - Falling off the stepladder.
  - Stepladder toppling (not fully opened).
- Falling away from the work (working from the side of the stepladder).
- Objects falling from height.

Safety precautions for stepladders:

- They should be maintained free from defects.
- They should be regularly inspected.
- They should not be painted.
- Retaining chains, ropes or bars (to prevent overspreading) should be of equal length and fully extended.
- They should be used on a firm, level base and be stable.
- Work should not be carried out from the top step.
- No more than one person at a time should use a stepladder.
- Over-reaching should be avoided.
- Do not work off the side of the stepladder, but facing the steps.

Correct and incorrect positioning of a stepladder

Trestles

A trestle work platform or scaffold is not normally suitable for high level work, and is only used for light work, e.g. painting, decorating. It provides a working platform which can be supported by A-frames, tripods, telescopic frames or split heads. A suitable and sufficient risk assessment is required prior to its use.

Hazards associated with using trestles:

- Falls from height.
- Trestle boards or platforms slipping.
- Equipment and objects falling from height.

Safety precautions for using trestles:

- Trestles should be maintained free from defects, regularly inspected, not be painted and be used on a firm, level base.
- Platforms based on trestles should be fully boarded, adequately supported and be provided with edge protection (toe boards, guardrails).
- Where appropriate, access to platforms should be provided (stepladder or ladder) and only one tier permitted when folding supports are used.

Staging Platforms

These have a number of different uses on access structures, e.g. on scaffolding, wooden trestles and mobile tower access, and are known under the name of Youngman's boards/platforms. They are also adaptable for use with guardrails and are available up to six metres in length for variable loads and situations.

Leading Edge Protection Systems

This subject is discussed thoroughly later in this element.

Mobile Elevating Work Platforms (MEWPs) and Mast Climbing Work Platform (MCWP)

Mobile Elevating Work Platforms (MEWPs) are motorised vehicles or trailers with powered extending arms supporting a work cradle. There are many different types of MEWP.

MEWP

Hazards associated with the use of MEWPs:

- Falls from the working platform.
- Objects falling from the platform.
- Collapse of the MEWP.
- Overturn (toppling) of the MEWP.
- Instability on poor terrain/crossing uneven ground.
- Contact with live cables or overhead services.
- Unauthorised use.

TOPIC FOCUS

Safety precautions for use of MEWPs:

- Select the type of MEWP to suit the terrain it will cross/work on.
- Site the MEWP on firm, stable ground.
- Ensure clearance from obstructions around and overhead when operating.
- Place barriers to prevent the MEWP being struck by vehicles or mobile plant.
- Same barriers keep people from beneath the working platform (cradle).
- Cradle to have guardrails; safety harnesses to be worn as additional precaution.
- Controls of the MEWP should be inside the cradle so the operator has some control while at height.
- MEWP should not be moved with the cradle raised unless designed to do so.
- Must not be overloaded.
- Must be inspected (six-monthly) as an item of lifting equipment designed to carry people.
- Restrict use to trained, authorised operators.

Mast climbing work platforms are also known as mast climbers, and are becoming more and more common on construction sites. Mast climbers have a power-driven work platform that climbs a vertical tower, allowing them to reach much higher and carry greater loads than traditional scaffolds.

The most common problems are:

- Falling from the platform. For instance, unguarded ends or removed guards.
- Climbing from the platform to a building opening.
- Overloading the platform.
- Instability of the mast climber during dismantling.
- Equipment failure.

Other Techniques

Boatswain's Chair

A boatswain's chair can be used for light, short-term work where there is no other practical means of access. The chair consists of a seat, a suspension point and means for carrying tools. The user should be attached to the chair by a harness to prevent falls. Control of descent is by the user, based on the same techniques as abseiling, although there should not be a single suspension point.

Boatswain's chairs being used for painting

TOPIC FOCUS

Installation of a boatswain's chair:

- To be installed and tested by a competent person.
- Work to be planned and supervised (as required by **WAH Regulations**).
- Chair to have two separate rope systems (main and security).
- Fall arrest harness to be attached to security rope.
- Tools should be secured in pouches or on tool lanyards.
- Means of communication provided for person in chair to ground operatives.
- Safety precautions in place to protect those working below.
- Rescue plan to be in place.
- Consider weather conditions before use.
- Safety equipment to be inspected daily.
- Chair and ropes examined every six months (SWL to be marked).

Cradles (Including Suspension from Cranes)

Suspended access equipment (cradles) are lowered into position from above. They can be either a fixed access structure built into the roof of a building, or a cradle designed to be positioned (raised and lowered) by a crane. They should be fully guarded with guardrails and toe boards to provide a safe working platform:

- Cradles must be thoroughly examined every six months by a competent person.
- An exclusion zone should be established below them, and operatives should wear harnesses inside the cradle for extra safety.
- Weather conditions can pose problems in the use of cradles, particularly wind and lightning.
- Cradles may also be used to deliver materials or people to working places, usually suspended by a crane.

Rope Access

Rope access (abseiling) can be used when access is not possible from a working platform:

- Users must be adequately trained and competent, and the equipment must be erected and inspected prior to use by a competent person.

- As such work requires two separately anchored lines (a working line for access, egress and support, and a safety line) both should have separate suspension points so that safe descent is not dependent on one rope.
- The user should wear a harness that is attached to both the working line and safety line, unless the use of a second rope would mean a higher risk to the person using the equipment. In this case, other appropriate safety measures should be used. The safety line should have a mobile fall protection system that is connected to and travels with the user.
- Tools or equipment used should be suspended by suitable ropes or chains to avoid being dropped. The area beneath rope access work must be protected by fans, covered walkways or tunnels to avoid any injuries.

How Fall Arrest Equipment is Used and its Aims

DEFINITION

FALL ARREST EQUIPMENT

Designed to catch you if you do fall.

Where all previous preventive methods are not practicable, or where additional safety measures are required, then fall arrest equipment should be used. This is intended to minimise the effect of a fall once it has happened, and is used as part of a safe system of work.

The provision of fall arrest equipment is subject to the **Personal Protective Equipment at Work Regulations 1992** and the **WAH Regulations**.

Fall arrest comes in two main forms:

- **Personal protective systems** - e.g. fall arrest harnesses.
- **Collective protection systems** - e.g. safety nets or air bags.

Harnesses

Personal fall arrest systems consist of a **full-body harness** connected to one or two lanyards, which are connected to a wire rope, fixed anchor point or inertia reel device:

- They should only be used by trained workers.
- All fall arrest equipment should be regularly inspected, maintained and cleaned.
- A higher level of supervision is required when these are used.
- The fall arrest equipment must be suitably anchored (to a suitable point strong enough to withstand the load from a fall), and as high as possible, preferably above the user and never below the level of their feet.
- Adequate clearance is required below the work level for safe operation.

Personal fall arrest equipment

Safety Nets

Safety nets are a **collective protection system**, for arresting falls of persons from height. They can be used on a variety of structures, e.g. towers, piers, bridges, chimneys, over motorways, or even across floor edges or gaps in walls using appropriate anchorage points.

Safety nets should be:

- Fixed to secure anchor points strong enough to support the net and the load of a person (or persons) falling into it.
- Fixed as close below the fall areas as possible, and never more than two metres below.
- Stored under cover in dry, clean conditions and protected from the effects of inclement weather, strong sunlight and contamination from oils, cement and other strong chemicals.
- Inspected before each use and after any fall incident for tears, holes or abrasion, especially at the anchor-attachment points. Such damage may be caused by people walking on the nets, dropped materials, tools or equipment, or friction against parts of the structure.
- Installed, inspected and examined at regular intervals according to manufacturers' specifications by a trained, certificated, competent person.

Soft Landing Systems

Cushioned reusable bags ('bean bags') or inflatable (low pressure) air bags provide a yielding surface onto which to fall.

The bags must be positioned correctly before work commences. They should be clipped together to ensure that they do not separate on impact. The bags should also be checked when re-positioned or after a fall has occurred to ensure that they have not been damaged in any way, and that air bag pressure has been maintained.

It is also important to ensure that other control measures are in place to prevent a fall from occurring in the first place. The bag system is a last resort and should not be relied on as the one and only control measure. It is important to remember that a fall even onto a 'safe' place could be very stressful and traumatic for an individual.

Crash Decks

Horizontal lining forms such as crash decks or tensioned cable netting are erected to protect equipment or items below from damage by objects falling from the work being carried out above. They are often semi-permanent structures erected over machinery to protect it in the event of materials falling, e.g. in the event of a roof collapse.

In public areas, it may be necessary to use pedestrian tunnels or properly constructed false ceilings or crash decks to protect pedestrians below from falling materials during work over occupied areas.

When scaffolding operations are in progress the public must be excluded from both the area of work and a sufficient area around it. It should be possible to incorporate fans, crash decks and tunnels as early as possible into a scaffold in the early stages of construction.

Tunnels can be constructed under scaffolding to protect the public

Emergency Procedures (Including Rescue)

Emergency procedures need to be developed for reasonably foreseeable events where workers might become trapped or suspended in their fall arrest equipment. Workers are placed at considerable risk if they become unconscious while suspended in a harness.

This step is often neglected when work at height is being planned.

Employers should consider the following:

- Rescue provision - the requirement to provide rescue provision.
- Adequate rescue equipment - the requirement to provide suitable and adequate rescue equipment.

The method of rescue may well be simple, such as putting a ladder up to a net and allowing a fallen person to climb down. Where harnesses are used, MEWPs may be required or proprietary rescue equipment employed.

Because rescue operations are often carried out under extreme pressure, consideration should be given to all aspects of the rescue process. It is worth considering the type of equipment required, the demands placed on the rescuer, the training that the rescue persons will require to carry out a rescue and how effectiveness of the system as a whole can be maintained

Suspension Fainting the Use of Rope and Harness

There is a general lack of awareness of suspension fainting (also known as 'suspension syncope'). Following an evidence-based review recently completed by the HSE, the guidelines are now that standard first-aid procedures should be given for suspension fainting (often incorrectly referred to as suspension trauma) once rescue and recovery of the employee has been achieved.

TOPIC FOCUS

Suspension fainting:

- Caused by pooling of the body's blood volume into the lower extremities.
- Can lead to changes in blood pressure, an increase in stress on the heart, and a reduction of oxygen supply to the brain.
- The longer a person is suspended in a harness while unconscious (without moving), the greater the chances of harm and the more serious it is likely to be. It can be fatal after only a short time.
- An unconscious person should be removed from upright suspension within ten minutes.
- A conscious casualty should be encouraged to exercise their legs to circulate the blood - this will reduce the harm.
- Medical services are to be immediately alerted in fainting cases, especially where the casualty is unconscious (kidney damage may result).

Roof Work

Certain precautions need to be taken during roof work to ensure the protection of:

- **Workers** - those actually working on a roof, such as:
  - Prevention of slips, trips and falls.
  - Safe places to walk.
  - Safe means of access and egress.
- **Occupants of buildings**, such as:
  - Information on the risks and precautions to be observed.
  - Clearance of areas beneath the work, barriers and signs, etc.

- **Members of the public**, such as:
  - Precautions to prevent people from being struck by falling materials/objects.
  - Barriers at ground level as well as at height.
  - Children must be protected and prevented from gaining access onto roofs.

Means of Access

Roof access requires careful planning as work progresses. This includes the means of transporting tools and materials to and from the roof, including the removal of waste. A number of methods may be used:

- Independent scaffolding can provide safe access to roofs and to the edges of them, as well as material storage areas.
- Fixed scaffold towers enable safe access to roofs. The use of stairway designs in preference to ladder access allows materials/components/tools to be carried to the roof safely.
- Mobile scaffold towers avoid the risks associated with scaffold erection, but must be suitable for the type of ground conditions present.
- Ladders - important considerations are the:
  - Length of ladder, which may be heavy and awkward to manoeuvre.
  - Need to carry materials or tools to and from the job, remembering that two hands should be free when climbing a ladder.
  - Stability and security of the ladder in use.
  - Type of ladder used.
  - Ladder requires inspection by competent persons.
- Where roof access is achieved by a valley or parapet gutter and the adjacent roof is of fragile material, then suitable covers are required to prevent a fall through the fragile material.

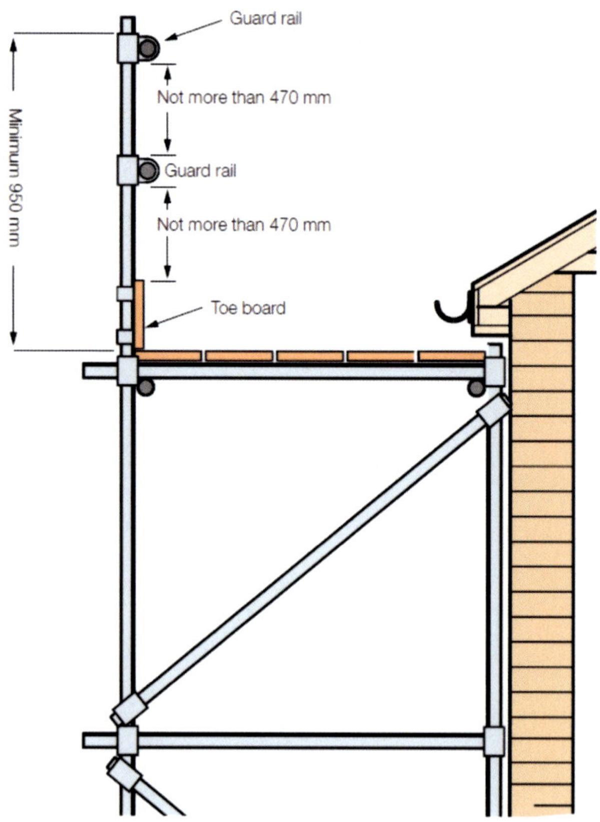

Dimensions of sloping roof edge protection (Source: HSG33 Health and safety in roof work, HSE, 2020 (www.hse.gov.uk/pubns/priced/hsg33.pdf))

Edge and Leading Edge Protection

Edge protection on roofs must be both high and strong enough to withstand a person either rolling or sliding down or falling off a roof structure, whether it is a flat or sloping roof. Protection may be permanent or temporary, e.g. guardrails or toe boards, depending on the roof structure.

Leading edges are unprotected edges (gaps) that are created and move as work progresses, e.g. where panels have been taken out in preparation for fitting new ones, or the edge of the line of panels as they are fitted (no roof beyond it). Working on a roof at the leading (working) edge gives rise to a number of risks which can be controlled in a number of ways:

- **Safety Nets**

  These need to be installed by a competent person as close as possible to the roof surface. They must be securely attached, of sufficient strength to withstand a person falling into the net and suitable for the work taking place. Note that they can also be used on refurbishment operations, e.g. near fragile roof lights.

- **Birdcage Scaffolding**

  Birdcage scaffolding is normally used for working inside factories, cinemas, churches or tall buildings. It can be used to gain access to ceilings and walls for plastering, painting or decorating, and to provide leading edge protection during roof refurbishments.

- **Safety Harnesses**

  Safety harnesses (as fall arrest systems) can be used which have running lines attached to a suitable anchorage point, mobile or fixed, e.g. a working platform. The harness needs to be fit for purpose, used in accordance with the manufacturer's instructions and the wearer suitably trained in its use (see earlier).

- **Trolley Systems**

  Temporary barriers such as trolley systems may be installed at the leading edge. Such systems require a safe system of work during installation, moving and dismantling, and must be able to be locked in position to prevent overturning. They also require safe access and must be able to move freely within the purlin (horizontal timbers resting on the principal rafters which support the common or subsidiary rafters or boards on which the roof is laid) design. Care is required that during their movement, additional safety is provided by the use of safety nets or harnesses and barriers at the end of the trolley.

  Attempting to free trolleys that have jammed can be dangerous. They are not suitable where design details such as hips, dormers, etc. do not allow adequate support over the full length of the trolley.

  They are not suitable for use where insufficient support is provided for the whole trolley length. Other systems can be protected by trolleys which limit the open area by providing a horizontal barrier attached to the trolley, which does not require a harness and is an alternative to the use of nets.

  Trolley systems are often not practicable and should only be used where there is no safer alternative

- **Other Systems**

  Other leading edge systems consist of soft matting which is used where safety nets are unable to be used (e.g. installation of precast slabs), or air-filled bag systems which are gaining popularity in the house-building sector.

Crawling Boards

These will be required on most sloping roofs and need to be:

- Fit for purpose.
- Of sufficient strength to support workers.
- At least three rafters long.
- Able to be secured or placed in such a way as to prevent any accidental movement.

At least two crawling boards should be used, one to support the worker while the other is moved to a new position.

On larger or more fragile roofs, crawling boards with side rails may be required, which give a worker extra support and anchorage for fall arrest equipment.

Other crawling-board systems include roof ladders that hook over the ridge, and timber battens.

Fall Arrest Equipment

Where all previous preventive methods are not practicable, or where additional safety measures are required, then fall arrest equipment (discussed earlier) should be used. This is intended to minimise the effect of a fall once it has happened, and is used as part of a safe system of work.

STUDY QUESTIONS

5. What safety features are used on scaffolding?
6. What are the main hazards of using ladders?
7. Outline the requirements for edge protection on working platforms.
8. In respect of scaffolding:
   (a) What is the difference between standards, ledgers and transoms?
   (b) What is the difference between tying and bracing?
9. State the safety precautions which need to be taken when mobile elevating work platforms are in use.
10. What is the angle at which ladders should be positioned?
11. What methods of roof access are there?
12. What is fall arrest equipment intended to do?

(Suggested Answers are at the end.)

Protection of Others

IN THIS SECTION...

- Danger areas are to be demarcated, segregated by barriers (fences, tapes, etc.) and tunnels (boarded walkways beneath work areas), and must be adequately signed to warn of the hazards.
- Marking and lighting are required to warn of and protect from the dangers on a construction site; lighting is particularly important for work in poor light and during the night where access is required.
- Sheeting, netting and fans are protective systems to prevent material and debris falling onto members of the public and other persons below work at height areas.
- Head protection should be provided for all persons who could be struck by falling or moving objects.

Introduction

Protection from falls from height or falling objects has to be provided for all persons, including site workers, contractors, visitors and members of the public.

Demarcation and Barriers

A suitable barrier a safe distance from the roof edge is used to demarcate between access routes and working areas. Normally, two metres from the edge is an accepted distance to prevent falls from the edge. Such barriers must be adequately supervised and suitably enforced to ensure they are not breached. The barriers themselves should be of sufficient strength and durability, adequately marked and with no visible gaps. The safe area similarly should have no unprotected holes, breaks or fragile material.

The internal area below any roof work and external areas around the edge of a building with roof work being carried out should be clearly segregated from vehicles, members of the public, children and anyone with impaired hearing or sight. This is achieved by the use of barriers - physical demarcation using fencing, mesh panels, cones or hazard tape. The higher the risk area, the more security and strength is required from the barrier.

Tunnels

Tunnels provide a shielded walkway through scaffolding and beneath work at height areas. They are protected on both sides (or just on the building side if along a main street footpath) and from above by boarding, sheeting or netting to prevent falling debris and other objects.

Signs

All demarcation areas, barriers, tunnels, etc. must be suitably signed to indicate the hazards and precautions to be taken, and to enforce their use.

Marking and Lighting

Marking of safe routes can be achieved by painted zones on the floor, on boarding and sheeting, or by the use of barrier tapes. All approaches (at all levels) to roof work should be signed.

Where demarcated areas are left overnight, adequate lighting is to be provided that illuminates any dim or shadowy areas. Dark areas (e.g. tunnels) may need 24-hour lighting. In addition to illuminating the danger zone, the signs should be highlighted as well.

Marking and lighting on site

Sheeting

Sheeting combined with a second layer of scaffold boards or even plywood sheets can help to protect the public from falling objects and materials. For example, hot bitumen can either drip or fall from roofs and so physical barriers below the actual work are required. However, careful consideration of the loading of the scaffolding is essential.

Netting

There are two main types of safety net:

- Personnel nets with 100mm mesh - these are intended to catch a person falling from above. The maximum recommended fall distance of persons is six metres.
- Material or debris protection nets with smaller mesh (12-19mm) - these are intended to protect persons below from falling objects.

Netting should be:

- Properly secured and slung efficiently.
- Higher at the outer edge than at the inner to allow access of any rescue craft.
- Erected as close as possible to the working level.

Fans

If there is any risk of pedestrians or vehicles below being hit, then the employer must erect a protective scaffold fan. Suitable and sufficient physical protection measures (of which the scaffold fan is one) need to be provided to prevent objects falling from scaffolding and protect the public if there is a risk of falling objects.

If masonry work is taking place or cradles are being operated, then a scaffold fan is also needed.

Fans or screens must be wide enough to provide complete protection to the public, and to prevent material of any kind from falling onto the public highway. (An illustration of fan protection was included earlier in this element.)

Head Protection

Work at height areas always have a risk of falling or moving objects, so are always safety helmet ('hard hat') areas. This requirement covers not only the construction workers, but all persons who need to be in the areas below work at height, including members of the public where demarcation cannot totally segregate them from the work at height activities.

For supervisory purposes, some larger sites will require different coloured helmets for identification of construction workers, contractors and visitors/members of the public on site.

The issue of hard hats is not a replacement for other controls designed to prevent objects from falling, e.g. toe boards or nets, or to protect people from falling objects, e.g. exclusion zones.

A Sikh who is wearing a turban is exempt from any requirement to wear head protection while working on or visiting a site where construction work is undertaken.

STUDY QUESTION

13. With regard to safety netting:
    (a) What are the two main types of safety netting?
    (b) What are the criteria for installing safety nets?

(Suggested Answer is at the end.)

Summary

This element has dealt with some of the hazards and controls associated with working at height.

In particular, this element has:

- Described the risk factors such as distance of fall, and features such as fragile roofs, roof lights, deterioration of materials, unprotected edges, poorly maintained access equipment, bad weather and falling materials.
- Outlined methods of avoiding working at height, and precautions to prevent falls and falling materials.
- Considered the importance of preparing emergency rescue plans and measures to be taken to reduce the consequences of a fall.
- Described the safe working procedures to be followed when erecting, using and dismantling scaffolding, and the design features of independent tied, putlog, fan and cantilevered scaffold and mobile scaffold towers.
- Described scaffold safety features, including sole boards, base plates, toe boards, guardrails, brick guards and debris netting.
- Outlined the requirements for gaining access onto scaffolding and the design of loading platforms, and scaffold hoists for persons and materials.
- Described the methods to ensure stability of scaffolding, including the effects of materials and weather.
- Outlined the hazards and precautions associated with the use of MEWPs, ladders, stepladders, trestles, staging platforms and leading edge protection systems.
- Outlined other access techniques: boatswain's chairs, suspended cradles and rope access; and the use of fall arrest equipment including harnesses, safety nets, crash decks and the requirement for rescue procedures (including for workers suffering suspension fainting).
- Described safety in roof work, including means of access, edge and leading edge protection, and the use of crawling boards and fall arrest equipment.
- Outlined the measures required to ensure the protection of other persons, including demarcation of danger areas, barriers, tunnels, signs, marking and lighting, sheeting, netting and fans, and the use of head protection.

Exam Skills

Question

Scenario

The construction site has been exposed to bad weather including heavy rain and high winds. A scaffolding contractor has arrived on site to erect an independent tied scaffold to enable bricklayers to work safely on the second level of the project. The site manager has asked you to liaise with the scaffolders during the erection of the scaffold to ensure the scaffolding is erected correctly and its design is going to be suitable for the purpose for which it is required.

Task: Scaffloding

Before meeting the scaffolding foreman, you decide to make some notes on aspects that need to be considered to ensure the scaffolding will be fit for purpose and maintained safely whilst in place.

(Total: 10 marks)

Approaching the Question

Now think about the steps you would take to answer this question:

Step 1 The first step is to **read the scenario carefully**. Note that the construction site has been exposed to bad weather (heavy rain and wind). Consideration of ground conditions and moving vehicles need to be made, as well as the load (people, materials and equipment) the scaffolding will be exposed to. .

Step 2 Now look at the **task** - prepare some notes. You will need to structure your approach, so think through the erection of the scaffolding structure in a logical sequence, e.g. ground upwards.

Step 3 Next, consider the **marks** available. In this task, there are 10 marks available, so it is expected that around 8 or 9 different pieces of information should be provided, as some pieces of information may gain more than 1 mark as they will require additional detail. Next, consider the **marks** available. In this task, there are 10 marks distributed between three parts. Tasks that are multi-part are often easier to answer because there are additional signposts in the question to keep you on track. In this task, you have to create some notes that are easy to understand, giving examples for each part that can aid understanding. You will need to provide around 10 or more different pieces of information (including examples) in total for this task. The headings will allow you to keep your responses separate – this will also help the examiner when marking!

Step 4 **Read the scenario and task** again to make sure you understand the requirements. (Re-read your study text if you need to.)

Step 5 The next stage is to **develop a plan** - there are various ways to do this. Remind yourself, first of all, that you need to be thinking about a note.

Suggested Answer Outline

The aspects that need to be considered to ensure safe erection of a scaffold are:

- Competency of scaffolders erecting the scaffold.
- Ground conditions.
- Condition of scaffolding poles and boards.
- Weight loading on the platform.
- Tie-in points.
- Level of bracing incorporated into the scaffold
- Proximity of mobile plant.
- Access and egress on to the scaffold.
- Inspection regime.

Now have a go at the question yourself.

Example of How the Question Could be Answered

To ensure the scaffold is erected and maintained correctly, a number of points would need to be established.

First, the personnel involved in its erection would have to be competent, so requiring proof of training records for those involved would be necessary.

Due to the weather conditions that have been encountered, ensuring the ground conditions are firm enough to withstand the loading imposed would have to be checked before erection.

Condition of the scaffolding components would have to be inspected, in particular scaffolding poles which might be bent and boards for cracks.

The design of the scaffold in relation to its purpose and weight loadings that will be imposed such as personnel, tools and materials. This also needs to consider dynamic loads such as wind. This in turn will decide on the level of tie-in points and where and how this will be done. To ensure the scaffold is stable, bracing will have to be incorporated into its design which will ensure structural robustness.

Consideration of vehicle traffic routes close to the scaffold may need to be diverted in order to ensure distance from the structure as well as incorporating physical barriers to prevent collision by mobile plant and vehicles.

Access and egress by personnel on to the scaffold would need to be established which could be by a temporary staircase or by ladders.

Identifying the number of staircases and ladders and their location would take into account the safe occupancy and to provide safe escape in the event of an emergency such as fire.

Finally, during the life of the scaffold structure, an inspection regime would have to be established at key intervals such after first erection, every 7 days, after inclement weather and after alterations or sudden shock loading (collision by a vehicle).

Reasons for Poor Marks Achieved by Exam Candidates

- Not expanding the answer beyond a few words as opposed to giving a sentence of explanation.
- Focussing on two or three areas and going into too much detail opposed to going wider covering all relevant points.

Unit NC1

Suggested Answers: Part 1

No Peeking!

Once you have worked your way through the study questions in this book, use the suggested answers on the following pages to find out where you went wrong (and what you got right), and as a resource to improve your knowledge and question-answering technique.

Element 1: The Foundations of Construction Health and Safety Management

Question 1

Financial impact affects:

(a) Lost production, repair of damage, replacement workers.

(b) Emergency services, hospitals.

Question 2

Risk areas of high-injury incidence include:

- Falls from height.
- Slips, trips and falls.
- Being struck by falling/moving objects.
- Manual handling.

(Note: only two were required.)

Question 3

The five duty holders are the client, the designers, the principal designer, the principal contractor and contractors.

Question 4

The HSE is notified of a project if the construction work in it will last longer than 30 working days and has more than 20 workers on the project. Or if it will involve more than 500 person days.

Question 5

CDM 2015 applies to all construction work in GB.

Question 6

The two documents that are required by the **CDM Regulations** are the construction phase health and safety plan and the health and safety file.

Question 7

Construction work falls into five separate stages, namely:

- **Concept and feasibility** - design and layout, construction methods involved.
- **Design and planning** - specifications required, final drawings, notification to authorities (unless exempt).
- **Tender/selection** - selection process for the Principal Contractor.
- **Construction phase** - planning, programming and preparing the construction work, and then starting, carrying out and managing the work involved.
- **Commissioning and handover** - all activities involved in ending the project and handing over to the client.

Question 8

Construction work includes building work; renovation; structural alteration; maintenance; decommissioning; civil engineering and works of engineering construction; and dismantling and demolition.

Question 9

The range of activities involved is varied, for example:

- Site clearance - removal of bricks, hardcore, scrap materials.
- Excavation, including earthworks, trenches, wells, shafts, tunnels or underground working.
- Loading, unloading, handling and storage of equipment and construction materials.
- Site movements of people, plant, machinery and vehicles.
- Use of access equipment, e.g. ladders and stepladders, working platforms, scaffolding, cradles, mobile platforms, crawling boards, gangways, trestles, suspended access equipment, etc.
- Fabrication, assembly or disassembly of parts of a structure.
- Decoration or redecoration following completion of work.
- Cleaning involving high-pressure water or corrosive substances.
- Installation, removal and maintenance of services (electricity, water, gas, compressed air, telecommunications, computer or similar services).
- Landscaping of grounds following completion of work or as part of a project.

Question 10

The following factors should be taken into account:

- Previous/current use of the site.
- History of the site.
- Area of the site and any restrictions placed on it.
- Topography.
- Ground conditions.
- Other, non-construction activities on site.
- Nature of the surroundings.
- Means of access.
- Presence of overhead and buried services.

Question 11

Restrictions on land for construction could include:

- National planning authorities placing planning or statutory restrictions.
- Local authorities having regulations, planning restrictions and building by-laws.
- Buildings listed for preservation.
- Government trade and industry regulations requiring permission for industrial development.

- Rights of way.
- Rights of support.
- Rights of light.
- Tunnels, mine workings and abandoned, active or proposed mineral rights.
- Ancient monuments (fortresses, tumuli, etc.) and burial grounds.
- Previous potentially contaminating uses of the site and adjacent areas.

(Note: only four were required.)

Question 12

Site security consists of:

- Traditional physical and electrical/electronic measures.
- Management and procedural measures.

For example:

- Perimeter fencing.
- Security guards on the premises.
- Locked gate.
- Means of securing plant and chemicals.
- CCTV and electronic intruder alarms.
- Roadside and security lighting.
- Safe viewing points.
- Visitor registration and sign-in procedures.
- Escorting persons while on site.
- Education, especially of children.

Question 13

Site rules should cover areas such as:

- Site access.
- Permit-to-work system.
- The wearing of PPE.
- Fire prevention rules.
- Welfare facilities.
- Alarms and emergency response.
- First-aid facilities.
- Accident reporting.
- Site transport precautions.

The actual rules will depend on the nature of the site and the type of industry where the construction work is being carried out.

Question 14

Five ways the public can be protected:

- Pedestrian tunnels and walkways with lighting.
- False ceilings and crash-decks.
- Sheeting on scaffolding.
- Wrapping protrusions (bolts, couplings, etc.).
- Barriers and signs around excavations.

Safe viewing points in safe areas and vision panels in solid hoardings.

Question 15

Precautions necessary to prevent accidents occurring with children on construction sites include:

- Security measures.
- Secure perimeter fence/hoarding which is appropriately signed.
- Removal of any access ladders to scaffolding, buildings, working platforms, etc.
- Providing secure and adequate trench supports.
- Immobilising plant and equipment, e.g. excavators, dumpers, vans.
- Isolating electrical equipment and ensuring the integrity of emergency supplies to the site.
- Reducing the height of any stacked materials.
- Filling/covering any holes.
- Blocking off debris chutes to prevent them being used as slides, i.e. by lids or covers.
- Locking away any equipment, chemicals, flammable materials in safe and secure storage areas.
- Securing storage tanks for oil, petrol, diesel.
- Visiting local schools and giving presentations on construction site dangers.

Question 16

Six out-of-hours security arrangements that might be used on a construction site?:

- CCTV - signage must be included warning of the use of CCTV.
- Lighting - suitable site and perimeter lighting.
- Security guards - to improve out-of-hours security.
- Storage compounds - locked and secured.
- Ladders and scaffolds - remove ladders or make access difficult.
- Excavations - barriers and covers.
- Falling objects - toe boards, brick guards and netting on scaffold.

(Note: only six were required.)

Question 17

Types of temporary works include (any four from):

- Formwork.
- Falsework.
- Scaffold.
- Temporary equipment foundations.
- Shoring in an excavation.
- Site hoardings and signage.
- Cofferdams and caissons.

Question 18

Employers must provide the following welfare facilities on a construction site:

- Drinking water.
- Sanitary conveniences.
- Washing facilities.
- Changing rooms and lockers.
- Rest and eating facilities.

Question 19

Issues that need to be considered when planning and managing a construction project are:

- Use of migrant workers.
- Temporary nature of construction activities.
- Time pressures from clients and contractors.
- Weather conditions.
- Workers' numeracy/literacy skills.
- Communicating with workers speaking different languages.

Element 2: Improving Health and Safety Culture and Assessing Risk

Question 1

The health and safety culture of an organisation is the way that everyone within the organisation thinks and feels about health and safety and how this translates into their behaviour. It is either positive or negative.

Question 2

An individual's peers exert influence through peer group pressure. This is the process by which social groups form in the workplace, group behaviour is established ('norms') and then social pressure is exerted to force individuals to comply with the group behaviour. There will usually be one or more group leaders who influence the group to a very high degree.

Question 3

Organisational factors, job factors and individual factors; collectively, these are the three 'human factors' that influence safety-related behaviour.

Question 4

The key job factors include:

- Task.
- Workload.
- Environment.
- Display and controls.
- Procedures.

Question 5

Perception can be defined as the way that a person interprets information detected by their senses.

Question 6

Written

Limitations	Merits
• Indirect. • Takes time to write. • May contain jargon and abbreviations. • Can be impersonal. • Message may be ambiguous. • Message may not be read by recipient. • Language barrier may exist. • Recipient may not be able to read. • Immediate feedback is not available. • Questions cannot be asked. • Recipient may have impaired vision.	• Permanent record. • Can be referred back to. • Can be written very carefully to avoid use of jargon, abbreviations and ambiguity. • Can be distributed to a wide audience relatively cheaply.

Verbal

Limitations	Merits
• Language barrier may exist. • Jargon may not be understood. • Strong accent or dialect may interfere. • Background noise may interfere. • Recipient may have poor hearing. • Message may be ambiguous. • Recipient may miss information. • Recipient may forget information. • No written record as proof. • Poor transmission quality if by telephone or PA system.	• Personal. • Quick. • Direct. • Allows for checking of understanding. • Allows for feedback to be given. • Allows for exchange of views. • Usually allows for additional information to be transmitted by means of tone of voice, facial expression and body language.

Question 7

Graphic symbols are predominantly in safety signs and hazard-warning labels. Sometimes used on posters.

Question 8

The first priority in induction training should be to set out what to do in the event of a fire or other major incident, and the general instructions and procedures to be followed for safe movement around the workplace.

Question 9

Whenever there is a significant change to the job which employees are to perform, a significant change to the processes they are to carry out, routinely as refresher training, after an accident, or near miss, after the introduction of new technology and when the relevant legislation changes.

Question 10

Employees can be involved by:

- Encouraging their participation in safety committees and other safety meetings.
- Asking for their suggestions for improvements (perhaps using a suggestions box or similar scheme).
- Involving them in the selection of PPE and other equipment.
- Providing them with hazard-spotting and defect reporting systems.
- Encouraging their participation in safety tours and inspections, audits, risk assessments, accident investigations and the development of procedures and safe systems of work.
- Assisting in the presentation of safety training, and the supervision/mentoring of new employees, particularly young persons.
- Designing or involvement in the selection of safety posters.

Question 11

Employers must consult employees directly, or through elected representatives (Representatives of Employee Safety) in a non-unionised workplace. Through Trade Union Safety Representatives in a unionised workplace.

Question 12

Trade union safety representatives have the right to:

- Carry out an inspection of the workplace (at least once every three months).
- Examine the causes of accidents.
- Examine and copy records and documents relating to health and safety.
- Receive information from HSE inspectors.
- Investigate complaints about health and safety.
- Make representations to the employer.
- Be consulted on health and safety matters.
- Time off with pay to perform their functions.
- Time off with pay for appropriate training.
- Reasonable facilities to perform their functions.

Question 13

Employers must establish a safety committee when requested to do so in writing by two or more TU safety representatives. The employer then has three months to set the committee up.

Question 14

Hazards will always exist in the workplace and usually it is not possible to eliminate them. Risk can be controlled and reduced. This is a central principle of health and safety management.

Question 15

The aim of risk assessment is to eliminate hazards or reduce risk to an acceptable level. The objectives are to prevent personal injury and ill health; to achieve legal compliance and to reduce the costs associated with losses.

Question 16

Techniques used for identifying hazards are: task analysis, analysis of incident data, examination of legislative requirements and associated guidance, examination of manufacturers' information.

Question 17

The five categories of health hazard are: physical, chemical, biological, ergonomic and psychological.

Question 18

The steps involved in risk assessment are: identify the hazards, decide who might be harmed and how, evaluate the risks and decide on precautions, record the findings and implement them, and review the assessment and update if necessary.

Question 19

Staff groups that require special consideration are: maintenance staff, cleaners, contractors, visitors, young workers, lone workers, new and expectant mothers and disabled staff.

Question 20

The factors used to evaluate risk are the likelihood of harm occurring and the severity of that harm.

Question 21

Residual risk is the level of risk remaining after the application of safety precautions.

Question 22

The general hierarchy of control:

- Elimination of the hazard.
- Substitution.
- Engineering controls.
- Administrative controls.
- Personal protective equipment.

Question 23

Factors that might trigger the review of a risk assessment include changes in legislation, a significant change in work practices and processes, installation of new machinery and equipment, new information becoming available on the hazards/risks, recurring accidents or patterns of ill health, enforcement action, results of monitoring/auditing, or employment of a category of personnel (e.g. disabled) not previously taken into account.

Question 24

Personal protective equipment should be used when it has not been possible to eliminate the hazard or reduce risk to acceptable levels by the use of engineering controls, working methods or administrative controls. PPE is a last resort.

Element 3: Managing Change and Procedures

Question 1

The impact of change is managed by:

- Communication and co-operation - the change must be effectively communicated.
- Risk assessment - as the nature of work changes, regular reviews of risk must take place.
- Appoint competent people - professional competent advice.
- Segregation of work areas - construction sites change rapidly, updates will be required
- Amendment of emergency procedures - fire exits and muster points may need to be repositioned.
- Welfare provision - numbers and types of workers will change as the activities change.

Question 2

A review system is important because regular reviews help to ensure the construction phase plan is kept on track.

Question 3

A safe system of work is a formal procedure which results from a systematic examination of the tasks of a work process in order to identify all the hazards and define methods of working which eliminate those hazards or minimise the risks associated with them.

Question 4

Involvement enables employees to gain a deeper understanding of hazards and risks, and of the way in which safe systems of work will minimise those risks. It also encourages ownership of key controls by the employees involved in their development.

Question 5

Technical (or engineering) controls are those which are applied directly to the hazard itself in order to minimise the risk (e.g. a machine guard). Procedural controls define the way in which work should be carried out in relation to the hazard (e.g. an operating instruction). Behavioural controls define how the individual operator or groups of workers must act in relation to the hazard (e.g. the use of an item of PPE).

Question 6

Instruction and training form a part of safe systems because only people who have been given appropriate training and instruction should be allowed to undertake the work.

Question 7

A permit to work is a formal document specifying the work to be done, hazards, and the precautions to be taken. Work can only start when safe procedures have been defined and put into place. The permit provides a clear written record, signed by a responsible manager or supervisor, that all foreseeable hazards have been considered and all the necessary actions have been taken. It should usually be in the possession of the person in charge of the operation before work can begin.

Question 8

The four key elements of a typical permit are issue, receipt, clearance/return to service, cancellation.

Question 9

The main objective of an emergency procedure is to ensure the safety and health of staff and others who might be affected by the emergency.

Question 10

Types of incident that may require development of an emergency response are: fire, bomb threat, spillage of a hazardous chemical, release of a toxic gas, outbreak of disease, severe weather or flooding, multiple casualty accident, terrorist/security incident.

Question 11

Things to consider for first aid provision would include:

- The general risk level of the workplace.
- The hazards present in the workplace.
- Accident history.
- The presence of vulnerable persons.
- The number of workers in the workplace.
- Work patterns and shift systems of workers.
- Provision for non-employees (e.g. shoppers, schoolchildren).
- The geographic location of the workplace.
- Distribution of the workforce (e.g. number of buildings on a site, number of floors, etc.).
- The physical size and spread of the workplace.
- The need to cover holidays and absence.
- Peripatetic employees (those who travel from place to place).

Question 12

The purpose of signs on an escape route is to direct occupants along routes.

Question 13

The main requirements for an escape route is to be available for every person in the workplace.

Question 14

An assembly point is where people gather once outside a building or away from a danger area. It will be in the open air and away from further danger from the fire. The assembly point differs from the refuge which is often provided inside the building in a place of comparative safety where people may wait a short time.

Question 15

The involved employee or, if they are not able to, a colleague, is usually responsible for initially reporting accidents and other safety-related incidents. In some circumstances, it might be the first aider who gave treatment. It might be the immediate line manager. Company policy will dictate.

Question 16

The purpose of analysing information about accidents is to identify immediate and root causes of accidents and to provide information about trends and patterns in workplace accidents.

Question 17

Depending on circumstances, the result of an individual accident investigation would be communicated to the casualty, their immediate manager and the local safety representative. It may also be necessary to inform other managers in the organisation, including senior management and the safety committee. For reportable incidents, the results might also be notified to the enforcement agency.

Question 18

The six main categories of reportable incidents under **RIDDOR** are: fatalities, specified injuries, over-seven-day injuries, hospitalisation of a non-worker, dangerous occurrences and reportable occupational diseases.

Question 19

Fatalities and specified injuries must be reported to the HSE Incident Contact Centre as soon as possible after the event by telephone or online form. This must be followed by full online notification within 10 days.

Immediate hospitalisation of a non-worker, occupational diseases and dangerous occurrences must be reported as soon as possible using the online form. This must be followed by full online notification within 10 days.

Over-seven-day injuries must be reported by online form within 15 days of the incident. There is no immediate notification requirement.

Question 20

Records of reportable incidents should be kept for at least three years.

Question 21

Occupational diseases that should be reported are carpal tunnel syndrome; severe cramp of the hand or forearm, occupational dermatitis, hand-arm vibration syndrome, occupational asthma, tendonitis or tenosynovitis of the hand or forearm, an occupational cancer, and any disease attributed to an occupational exposure to a biological agent.

Question 22

The types of information may include:

- Accident documentation about other, similar, incidents.
- Risk assessments relating to drilling work in that particular workplace.
- Inspection and maintenance records in respect of the particular equipment and guard, and others in the area.
- Other recent and relevant documentation relating to the use of guards, reporting of faults, employee history (including training records), etc.

Question 23

The following action should be taken:

- Make the area safe (in some circumstances this might need to be done first so there is no risk to those giving assistance).
- Ensure first aid and then medical treatment for the victim.
- Isolate the scene so evidence is not disturbed.
- Inform next of kin.
- Notify enforcing authority if relevant.
- Arrange any necessary counselling or support.
- Set up investigation team:
  - Collect evidence, including photographs, measurements, etc.
  - Take statements from witnesses.
  - Determine immediate and root causes.
  - Report, making recommendations to prevent recurrence.
- Inform safety representatives, and issue internal information.
- Advise insurers.
- Implement recommendations, revise work procedures and risk assessments.
- Monitor situation and review as necessary.

Element 4: Excavation

Question 1

The hazards of work in and around excavations include:

- Buried services.
- Falls of persons/equipment/material into the excavation.
- Collapse of sides.
- Collapse of adjacent structures.
- Water ingress.
- Use of cofferdams and caissons.
- Contaminated ground.
- Toxic and asphyxiating atmospheres.
- Mechanical hazards.

Question 2

Risk assessment factors to consider for excavations include:

- Depth.
- Soil type.
- Type of work.
- Use of mechanical equipment.
- Proximity of roadways/structures.
- Presence of the public.
- Weather.

Question 3

Controls that may be necessary during excavation work are:

- Identification and marking of buried services, and using safe digging methods to avoid contact with them.
- Methods of supporting excavations (e.g. steel sheets, support boxes).
- Safe means of access.
- Safe crossing points.
- Barriers.
- Lighting and warning signs.
- Safe storage of spoil.
- De-watering.
- Positioning and routeing of vehicles, plant and equipment.
- PPE.
- Additional controls will be required when working with contaminated ground, e.g. separate welfare facilities.

Question 4

Hazards that may arise during the excavation of contaminated ground:

- Digging will uncover buried materials or contaminants within the ground which are hazardous to health.
- The contaminants themselves can change over time due to bacterial or chemical action, which may also alter their properties. This may be the result of the decomposition of organic matter or from the dumping or spillage of hazardous substances.
- Certain contaminants are subject to specific legislation, namely asbestos, lead, anthrax, radioactive materials and buried explosives, and well established procedures need to be followed to deal with each of these situations.
- There may be radioactive hazards from the ground itself (probably as a result of previous occupancy).
- Toxic and asphyxiating atmospheres may be present, for example:
  - Flammable gases such as methane (marsh gas) and carbon monoxide from plant and machinery used in connection with excavation, including pumps involved in de-watering operations.
  - Toxic gases such as hydrogen sulphide.
  - Chemicals and metal compounds, either in containers or within the soil.

Question 5

No person is allowed to work in an excavation before it has been examined by a competent person.

The **CDM Regulations** require that a competent person shall inspect the working areas of an excavation at the start of each shift. Additional inspections will be required after any event likely to have affected the strength or stability of the excavation, or any part of it, e.g. flooding or other incidents such as an accidental fall of rock or earth or other material.

A report must be made and records kept of such inspections.

Question 6

(a) A confined space is any enclosed space where there is a reasonably foreseeable specified risk of any serious injury associated with it.

(b) Typical confined spaces found in construction work include trenches, sewers, manholes, tunnels, excavations, chambers, tanks, pits, cellars and unventilated rooms. Cofferdams and caissons by their enclosed nature may also be included.

Question 7

Hazards associated with confined spaces include:

- Exposure to toxic, explosive and oxygen-deficient atmospheres.
- Heat.
- Water.
- Free-flowing solids.
- Restricted space.

Question 8

Precautions that are necessary for safe entry into a confined space include:

- Avoidance where possible.
- Risk assessment.
- Planning.
- Permit-to-work procedures and implementation of their requirements.
- Training.
- Use of competent persons.
- Atmospheric testing.
- Means of access.
- PPE.

Question 9

Factors that might be involved in a safe system of work for confined space entry include:

- Adequate, competent supervision.
- Experienced and competent workers.
- Proper isolation of plant and equipment.
- Pre-cleaning and correct disposal of debris and sludge.
- Adequate lighting and ventilation.
- Atmospheric testing.
- Provision of specialist, non-sparking tools; breathing apparatus; and rescue harnesses.
- Emergency arrangements/plans.

Element 5: Demolition

Question 1

Three other methods of demolition are:

- Pre-weakening.
- Progressive demolition.
- Overturning by wire rope pulling.

Question 2

Transport-related accidents in construction work are due to:

- Poor, unstable and uneven ground.
- Poor control of vehicle movements.
- Use of unsuitable plant or vehicles.
- Unauthorised use of vehicles and plant by untrained operators.
- The potential proximity of private or road-going trade vehicles to heavy demolition plant.
- Vehicles left in unplanned areas, causing obstructions and obscuring sight lines.

(Note: only three were required.)

Question 3

The method statement and demolition plan should outline:

- How the presence of hazardous substances is to be identified.
- The means of disposing of such substances.
- The requirements for any protective equipment.

Question 4

To avoid the premature collapse of a building/structure during demolition, designers should consider:

- How structural stability can be maintained throughout the process and include relevant information with their designs in order to avoid such events taking place.
- That structures can become unsafe during the demolition process and temporary scaffolding/support may be required, e.g. for heavy overhanging cornices.
- The need to ensure that these temporary structures are able to support the loads required, so that overloading does not occur.
- That the removal of certain parts of the building or structure during demolition can result in other parts of the building becoming unsafe.
- That there are a number of types of temporary structural support for shoring, propping and facade retention including tube and fittings scaffolding; system scaffolding; screw jack propping; shoring systems; facade retention systems; and buttresses.
- That temporary structural supports should be used in accordance with the appropriate standards and should be sufficient and suitable for the work required.

- That temporary structural supports should be:
  - In position prior to the disturbance or removal of the existing supports.
  - Fit for purpose and founded on secure and effective footings.
  - Adequately laced and braced.
  - Capable of resisting dynamic loadings.
  - Checked both prior to and during demolition for effectiveness, including for the effects of vibration.

In addition:

- Scaffolding should normally be an independent tied scaffold. When completed, it should conform to the requirements of the statutory regulations and with any local authority requirements, e.g. inspected, adjusted as necessary and removed in a progressive and planned manner as the work proceeds, to ensure its integrity.
- Any shoring and propping should be adequate for its purpose, placed in position at the appropriate time and designed not to interfere with subsequent construction. The shoring should be checked for effectiveness as the demolition proceeds and should never be loaded in excess of the design limits.
- The temporary support system should not be weakened by removal of components until the facade in that location is fully supported either by the new structure or by an alternative, temporary structure.

Question 5

A number of steps need to be taken to avoid noise and provide protection for people working in, or living near, demolition sites.

- A noise assessment is required in accordance with the **Control of Noise at Work Regulations 2005,** which will identify all potentially noise-sensitive areas on the site.
- Monitor background noise prior to work beginning at any noise-sensitive areas.
- State actions taken to reduce noise levels (in the demolition method statement).
- Use plant and equipment with low operating noise levels, and mains-generated electricity in preference to diesel generators.
- Use solid-panelled fencing around the site instead of wire fences.
- Restrict deliveries to daytime working hours, i.e. not early morning or late at night.
- Display project contact details prominently in residential areas, to allow any nuisance issue to be addressed quickly.
- Post warning notices to warn people to keep away from the danger area.
- Fit appropriate silencing/soundproofing to all machinery and plant on site.
- Provide personnel working within the noise zone with, and ensure they use, adequate hearing protection.

Question 6

Protection against dust from demolition activities may include the following:

- A **COSHH** assessment.
- Maintaining a high standard of housekeeping, regularly cleaning and sweeping the site roads.
- Using techniques that produce low levels of dust; damping down.
- Designating vehicle routes to and from the site to limit the problem.

- Maintaining plant and equipment to a high standard, so that it is less likely to break down and will emit fewer pollutants, and to prevent spillage of oil and fuel to the environment.
- Using wheel-washing facilities at exits onto public traffic routes.
- Using water sprays as a dust suppressant, taking care as to the potential consequences of environmental contaminants or the proximity to electrical services.
- Using air movers or LEV where possible.
- Using covered wagons and skips and restricting the speed of vehicles.
- Using PPE as a last line of defence.
- Where asbestos might be present, take samples to verify its presence and carry out the necessary precautions, actions and required notification.
- Put precautions in place if lead is found to be present.

Question 7

To ensure the safe siting of plant and machinery:

- The plant and machinery should be properly sited within a compound and isolated or immobilised at the end of each shift.
- Sufficient and suitable parking areas should be provided for all vehicles using the site - including employees' and visitors' private vehicles.
- Vehicles should not be left in unplanned areas, otherwise the safe operation of the site may be jeopardised - for example, by obscuring sight lines.
- Safe areas for loading and unloading skips and collection vehicles should be provided.
- Prevent parking of vehicles near any scaffolding, stacks of materials, compressed gas or storage compounds containing toxic or flammable materials, especially transport carrying LPG.

Question 8

Environmental information relating to any underground tanks or potential chemical/biological hazards or the presence of contaminated land, along with the building's former use, should be supplied to the contractor prior to demolition.

Types of contaminated sites (including both ground and buildings) are numerous and can include waste disposal sites; gasworks, power stations; sewage treatment works; railway land including sidings and depots; coal-yards (contaminated by phenols); oil refineries, petrol stations and associated sites; metal mines, smelters, electroplating works, refineries and foundries; chemical works; Industrial Revolution era workshops and warehouses; slaughter yards and abattoirs; fish and meat markets; tanneries; paper and printing works; wood preservative treatment sites; hospitals, including isolation and quarantine premises; and buildings with animal hair used as insulation in roofs, etc. (anthrax).

Protection is required from noise, dust, waste and asbestos. Any particular problems that might arise from these should be taken account of in the risk assessment and detailed safety precautions should be put in place.

All the information should be included in the health and safety plan for the demolition project.

Question 9

Information to be considered in a pre-demolition survey:

- The presence (or absence) of asbestos; of what types and in what locations.
- The identification of key structural elements including pre- and post-tensioned components.
- The identification of the location and types of services.
- The identification, significance and extent of any dilapidation of the structure.
- A review of the drawings and structural calculations, and the existing health and safety file relating to the structure.
- A review of all structural alterations carried out on the structure in the past.

Element 6: Mobile Plant and Vehicles

Question 1

Reasons for falling objects may include:

- Unstable access equipment (which can collapse or overturn).
- Poor housekeeping and incorrect methods of getting materials, equipment or tools from ground level to the working area or vice-versa.
- Adverse weather conditions, e.g. high winds.
- Workers accidentally dropping things.
- Poorly secured or stacked objects.
- Wear and tear.
- Unprotected edges on scaffolding.

(Note: only four were required.)

Question 2

The three main types of hazard caused by vehicle operations are loss of control; overturning; and collisions with other vehicles, pedestrians or fixed objects.

Question 3

The loss of control of a vehicle may be caused by driver error, mechanical failure or environmental conditions.

Question 4

Several factors may contribute to overturning, including:

- Speed of travel.
- Steepness of the slope - travelling across a slope.
- Height of the vehicle (including its load).
- Stability of the load.
- High tyre pressure.
- Any external longitudinal pressure, e.g. wind or colliding objects.
- Presence and size of any bumps or holes in the surface.

Question 5

Adverse environmental conditions that may contribute to vehicle accidents include:

- Poor weather.
- Poor lighting.
- Poor sight lines.
- Poor ground surface.
- Congestion.
- Gradients and changes of level.

Question 6

Visibility can be improved by ensuring adequate lighting is provided for the traffic route (and its surroundings), and that sight lines are optimised.

Question 7

Management of vehicle operations and movements encompasses the following measures:

- Reversing controls.
- Speed limits.
- Vehicle parking.
- Signs and markings.
- Signalling.
- Loading and unloading procedures.
- Specific rules for particular sites.
- General rules for ensuring vehicles do not become hazards.

Question 8

Traffic routes can be maintained to ensure vehicle safety by:

- Potholes should not be allowed to develop.
- Loose materials should be cleared regularly.
- Snow clearing and gritting may be necessary.
- Temporary obstructions (including goods, etc. fallen from vehicles) should be cleared as soon as possible.
- Road-sweeping machines should be used to clean the surfaces and maintain them in good condition.

Question 9

The three main types of training available to drivers is general training, specific job training and refresher training.

Question 10

Factors associated with driving at work that can increase the risk of a traffic accident include:

- Distance.
- Driving hours.
- Work schedules.
- Weather conditions.

(Note: only three were required.)

Question 11

To ensure work-related driving is effectively managed:

- Ensuring the policy is in place and understood by all drivers.
- Ensuring all management responsibilities are realised and achieved.
- Putting in place an organisational structure to manage the risks.
- Having appropriate safety systems in place.
- Monitoring the road safety performance.

(Note: only four were required.)

Question 12

The main factors to consider when managing work-related road safety include:

- Safe driving policy.
- Systems to manage work-related road safety.
- Monitoring road safety performance.
- Organisation and structure.
- Legal responsibilities of individuals on the public highways.

(Note: only three were required.)

Element 7: Working at Height

Question 1

The **Work at Height Regulations 2005** call for a hierarchy of measures where there is a risk of a fall liable to cause personal injury, namely:

- Avoid work at height.
- Use work equipment or other measures to prevent falls where work at height cannot be avoided.
- Use work equipment or other measures to minimise the distance and consequences of a fall where the risk of a fall cannot be eliminated.

Question 2

The safe method of working on fragile roofs is by the use of roof ladders or crawling boards. These are laid across the surface, supported by underlying and load-bearing roof members, and distribute the load of the worker over a wider area, enabling the roof structure to sustain the load safely. Roof ladders also provide a good foot-and-hand hold for the worker.

Fall arrest equipment should be worn when shown to be required by a risk assessment.

Question 3

Commonly, falls occur:

- From the edge of a completed roof.
- From a roof where work is being completed.
- Through openings or gaps.
- Through fragile materials making up part of the roof.

Question 4

To prevent materials from falling in the first place, the following control measures should be used:

- Not stacking materials near edges, and particularly unprotected edges.
- Close boarding of working platforms to minimise the gaps between scaffold boards, or placing sheeting over the boards so that material cannot fall through.
- Avoiding carrying materials up or down ladders, etc. by using hoists and chutes to move materials.
- Using physical safeguards such as toeboards and brick guards.

Question 5

Safety features include sole boards, base plates, toe boards, guardrails, boarding, brick guards and debris netting and waste chutes.

Question 6

The main hazards of using ladders are:

- Not being tied or not resting on firm ground, which may cause the ladder to tip.
- Poor storage and maintenance allowing the ladder to rot or warp, which may cause the rungs to break.
- Objects falling from height.
- Contact with live overheads.

Question 7

Detailed requirements for edge protection are:

- Top guardrail to be at least 950mm above the working platform.
- Intermediate rails so no unprotected gap of more than 470mm between toeboard and guardrail.
- Toe board must extend to at least 150mm above the working platform.

Question 8

In respect of scaffolding:

(a) Standards are the vertical tubes (the uprights), ledgers are the horizontal tubes running parallel to the face of the building and transoms are the tubes spanning across ledgers to secure a scaffold transversely.

(b) Tying secures the scaffolding to the building, whereas bracing is used to stiffen the framework.

Question 9

Safety precautions for use of MEWPs:

- Select the type of MEWP to suit the terrain it will cross/work on.
- Site the MEWP on firm, stable ground.
- Ensure clearance from obstructions around and overhead when operating.
- Place barriers to prevent the MEWP being struck by vehicles or mobile plant.
- Same barriers keep people from beneath the working platform (cradle).
- Cradle to have guardrails; safety harnesses to be worn as additional precaution.
- Controls of the MEWP should be inside the cradle so the operator has some control while at height.
- MEWP should not be moved with the cradle raised unless designed to do so.
- Must not be overloaded.
- Must be inspected (six-monthly) as an item of lifting equipment designed to carry people.
- Restrict use to trained, authorised operators.

Question 10

Ladders should be positioned at an angle of 75° (4:1).

Question 11

Roof access methods include: independent scaffolding, fixed scaffold towers, mobile scaffold towers, ladders, or by a valley or parapet gutter.

Question 12

Fall arrest equipment is intended to minimise the effect of a fall once it has happened and is used as part of a safe system of work. The provision of fall arrest equipment is subject to the **Personal Protective Equipment at Work Regulations 1992** and the **Work at Height Regulations 2005**.

Question 13

(a) The two main types of safety netting are personnel nets (100mm mesh) to catch persons falling from above - maximum fall distance six metres; and material or debris netting (12-19mm mesh) to catch falling objects.

(b) Safety nets should be:

- Properly secured and slung efficiently.
- Higher at the outer edge than at the inner to allow access of any rescue craft.
- Erected as close as possible to the working level.